RUSSIAN COMBAT METHODS IN WORLD WAR II

PREFACE

This pamphlet was prepared by a committee of former German officers at the EUCOM Historical Division Interrogation Enclosure, Neustadt, Germany, in late 1947 and early 1948. All of these officers had extensive experience on the eastern front during the period 1941–45. The principal author, for example, commanded in succession a panzer division, a corps, a panzer army, and an army group.

The reader is reminded that publications in the GERMAN REPORT SERIES were written by Germans from the German point of view. For instance, the "Introduction" and "Conclusions" to *Russian Combat Methods in World War II* present the views of the German author without interpretation by American personnel. Throughout this pamphlet, Russian combat methods are evaluated in terms of German combat doctrine, and Russian staff methods are compared to those of the German General Staff. Organization, equipment, and procedures of the German and Russian Armies differed considerably from those of the United States Army. Tactical examples in the text have been carefully dated, and an effort has been made to indicate the progress of the Russian Army in overcoming the weaknesses noted in the early stages of the war.

In the preparation of this revised edition, the German text has been retranslated, and certain changes in typography and chapter titles have been made to improve clarity and facilitate its use. The revised edition is considered to be just as reliable and sound as the text prepared by the German committee. The authors' prejudices and defects, whatever they may be, find the same expression in the following translation as they do in the original German.

CONTENTS

MAPS

(In sequence inside back cover)

This pamphlet supersedes MS #T-22, "Peculiarities of Russian Warfare," published by the Historical Division, Special Staff, U. S. Army, in June 1949.

PART ONE

INTRODUCTION

The only written material available for the preparation of this manuscript consisted of a few memoranda in diary form and similar notes of a personal nature. *Russian Combat Methods in World War II* is therefore based to a preponderant degree on personal recollections and on material furnished by a small group of former German commanders who had special experience in the Eastern Campaign. For that very reason, it cannot lay claim to completeness. This report is limited to a description of the characteristic traits of the Russian soldier, and their influence on the conduct of battle. The political, economic, and social conditions of the country, although influential factors, could only be touched upon. Detailed treatment of climate and terrain—indispensable to an understanding of Russian methods of warfare—has been omitted intentionally since those subjects are discussed in other manuscripts.

Russian combat methods have more and more become a topic of vital concern. Propaganda and legend already have obscured the facts. The most nearly correct appraisal will be arrived at by knowing the peculiarities of the Russian territory and its inhabitants, and by analyzing and accurately evaluating the sources from which they derive their strength. There is no better method than a study of World War II, the struggle in which the characteristics of country and people were thrown into bold relief. Although the passage of time may have diminished the validity of these experiences, they nevertheless remain the soundest basis for an evaluation. The war potential of the Soviet Union may be subject to change; no doubt it has increased during the last few years and will increase further, at least until the end of the current Five Year Plan. The very latest implements of war are known to have been further developed and produced in quantity, and new offensive and defensive weapons perfected. Technological advances will alter the external aspects of warfare, but the character and peculiarities of the Russian soldier and his particular methods of fighting remain unaffected by such innovations. Nor will the characteristics of Russian topography change during the next few years. In these decisively important aspects, therefore, the German experiences of World War II remain fully valid.

PART TWO

THE RUSSIAN SOLDIER AND RUSSIAN CONDUCT OF BATTLE

Chapter 1

Peculiarities of the Russian Soldier

It is possible to predict from experience how virtually every soldier of the western world will behave in a given situation—but not the Russian. The characteristics of this semi-Asiatic, like those of his vast country, are strange and contradictory. During the last war there were units which one day repulsed a strong German attack with exemplary bravery, and on the next folded up completely. There were others which one day lost their nerve when the first shell exploded, and on the next allowed themselves, man by man, literally to be cut to pieces. The Russian is generally impervious to crises, but he can also be very sensitive to them. Generally, he has no fear of a threat to his flanks, but at times he can be most touchy about flanks. He disregards many of the old established rules of tactics, but clings obstinately to his own methods.

The key to this odd behavior can be found in the native character of the Russian soldier who, as a fighter, possesses neither the judgment nor the ability to think independently. He is subject to moods which to a Westerner are incomprehensible; he acts by instinct. As a soldier, the Russian is primitive and unassuming, innately brave but morosely passive when in a group. These traits make him in many respects an adversary superior to the self-confident and more demanding soldiers of other armies. Such opponents, however, can and must, by their physical and mental qualities, achieve not only equality, but also the superiority necessary to defeat the Russian soldier.

Disregard for human beings and contempt of death are other characteristics of the Russian soldier. He will climb with complete indifference and cold-bloodedness over the bodies of hundreds of fallen comrades, in order to take up the attack on the same spot. With the same apathy he will work all day burying his dead comrades after a battle. He looks toward his own death with the same resignation. Even severe wounds impress him comparatively little. For instance, a Russian, sitting upright at the side of the street, in spite of the fact that both lower legs were shot away asked with a friendly smile for a cigarette. He endures cold and heat, hunger and thirst, dampness and mud, sickness and vermin, with equanimity. Because of his simple and primitive nature, all sorts of hardships bring him but few emotional reactions. His emotions run the gamut from animal ferocity to the utmost kindliness; odious and cruel in a group, he can be friendly and ready to help as an individual.

In the attack the Russian fought unto death. Despite most thorough German defensive measures he would continue to go forward, completely disregarding losses. He was generally not subject to panic. For example, in the break-through of the fortifications before Bryansk in October 1941, Russian bunkers, which had long since been bypassed and which for days lay far behind the front, continued to be held when every hope of relief had vanished. Following the German crossing of the Bug in July 1941, the fortifications which had originally been cleared of the enemy by the 167th Infantry Division were reoccupied a few days later by groups of Russian stragglers, and subsequently had to be painstakingly retaken by a division which followed in the rear. An underground room in the heart of the citadel of Brest Litovsk held out for many days against a German division in spite of the employment of the heaviest fire power.

The sum of these most diverse characteristics makes the Russian a superior soldier who, under the direction of understanding leadership, becomes a dangerous opponent. It would be a serious error to underestimate the Russian soldier, even though he does not quite fit the pattern of modern warfare and the educated fighting man. The strength of the Western soldier is conscious action, controlled by his own mind. Neither this action on his own, nor the consciousness which accompanies the action, is part of the mental make-up of the Russian. But the fact must not be ignored that a change is taking place also in this respect.

The difference between the Russian units in World War I and those in World War II is considerable. Whereas in the earlier war the Russian Army was a more or less amorphous mass, immovable and without individuality, the spiritual awakening through communism showed itself clearly in the last war. In contrast to the situation at the time of World War I, the number of illiterates was small. The Russian masses had acquired individuality, or at least were well on the way to acquiring it. The Russian is beginning to become a perceptive human being, and hence a soldier who is able to stand on his own feet. The number of good noncommissioned officers was still not large in World War II and the Russian masses had not yet overcome their sluggishness. But the awakening of the Russian people cannot be far off. Whether this will work to the advantage or disadvantage of their soldierly qualities cannot yet be determined. For along with awareness flourish criticism and obstinacy. The arbitrary employment of masses resigned to their fate may become more difficult, and the basis of the typically Russian method of waging war may be lost. The force bringing about this change is communism, or more precisely, a spiritual awakening of the people directed by a rigidly centralized state. The Russian is fundamentally nonpolitical; at least that is true for the rural population, which supplies the majority of soldiers. He is not an active Communist, not a political zealot. But he is—and here one notes a decisive change—a conscious Russian who fights only in rare instances for political ideals, but always for his Fatherland.

In judging the basic qualities of the Russian it should be added that by nature he is brave, as he has well demonstrated in his history. In 1807 it was the Russian soldier who for the first time made a stand against Napoleon after his victorious march through Europe—a stand which may be called almost epic.

In line with this awakening, another determining factor has been introduced into the Red Army by the political commissar—unqualified obedience. Carried out to utter finality, it has made a raw mass of men a first-rate fighting machine. Systematic training, drill, disregard for one's own life, the natural inclination of the Russian soldier to uncompromising compliance and, not the least of all, the real disciplinary powers available to the commissar, are the foundations of this iron obedience. In this connection, it must be remembered that Russia is an autocratically ruled state—an absolute dictatorship demanding and compelling the complete subordination of the individual. That blind obedience of the masses, the mainspring of the Red Army, is the triumph of communism and the explanation of its military successes.

In addition to the simplicity which is revealed in his limited household needs and his primitive mode of living, the Russian soldier has close kinship with nature. It is no exaggeration to say that the Russian soldier is unaffected by season and terrain. This immunity gave him a decisive advantage over the Germans, especially in Russian territory where season, temperature, and terrain play a decisive role.

The problem of providing for the individual soldier in the Russian Army is of secondary importance, because the Russian soldier requires only very few provisions for his own use. The field kitchen, a sacred institution to other troops, is to the Russian soldier a pleasant surprise when it is available, but can be dispensed with for days and weeks without undue hardship.

During the winter campaign of 1941, a Russian regiment was surrounded in the woods along the Volkhov and, because of German weakness, had to be starved out. After 1 week, reconnaissance patrols met with the same resistance as on the first day; after another week only a few prisoners were taken, the majority having fought their way through to their own troops in spite of close encirclement. According to the prisoners, the Russians subsisted during those weeks on a few pieces of frozen bread, leaves and pine needles which they chewed, and some cigarettes. It had never occurred to anyone to throw in the sponge because of hunger, and the cold (−30° F.) had not affected them.

The kinship with nature, which the Russians have retained to a greater degree than the other peoples of Europe, is also responsible for the ability of the Russian soldier to adapt himself to terrain features, and actually to merge with them. He is a master of camouflage, entrenchment, and defense construction. With great speed he disappears into the earth, digging in with unfailing instinct so as to utilize the terrain to make his fortifications very difficult to discover. When the Russian has dug himself into his native soil and has molded himself into the landscape, he is a doubly dangerous opponent.

The utmost caution is required when passing through unknown terrain. Even long and searching observation often does not reveal the excellently camouflaged Russian. Frequently, German reconnaissance patrols passed through the immediate vicinity of Russian positions or individual riflemen without noticing them, and were then taken under fire from behind. Caution must be doubled in wooded areas. In such areas the Russians often disappear without a trace, and must be driven out individually, Indian fashion. Here, sniping from trees was particularly favored by the Russians as a method of fighting.

The industrialization of the country, carried out in a comparatively short period of time, has made available to the Red Army a large number of industrial workers with full command of technical skills. The Russian has mastered all new weapons and fighting equipment, all the requirements of machine warfare, with amazing rapidity. Soldiers trained in technical subjects were carefully distributed through the ranks where they taught the necessary rudiments to their duller urban comrades, and to those who came from rural areas. The technical skill of the Russian was especially notable in the field of signal communications. The longer the war lasted, the better the Russians became at handling this type of equipment. Their communications improved steadily, and with noteworthy skill the Russians soon made themselves familiar also with German signal communications. Monitoring and tuning in, jamming and deception, all were arts which they understood very well. Whereas in World War I the telephone was still magic to the average Russian, in World War II he regarded the complicated radio as an amusing toy. In the field of signal communications, he also maintained his iron discipline, thereby propounding many a riddle for German signal intelligence.

In contrast to the good side of the Russian soldier there were bad military aspects of equal significance. To the Germans, it was one of the imponderables about each Russian unit whether the good or bad would predominate. There still remained an appreciable residue of dullness, inflexibility, and apathy which had not yet been overcome, and which probably will not be overcome in the near future.

The unpredictability of the mood of the Russian soldier and his pronounced herd instinct at times brought on sudden panic in individual units. As inexplicable as the fanatic resistance of some units, was the mystery of their mass flights, or sudden wholesale surrender. The reason may have been an imperceptible fluctuation in morale. Its effect could not be counteracted by any commissar.

His emotions drive the Russian into the herd, which gives him strength and courage. The individual fighter created by modern warfare is rare among the Russians. Most of the time a Russian who has to stand on his own feet does not know what to do. During the war, however, this serious weakness was compensated for by the large mass of men available.

A word about the craftiness of the Russian. He seldom employed large-scale ruses. The usual tricks, such as feigning the existence of troops by increased fire and other means, were just as common with the Russians as with all armies. They seldom carried out feint attacks. The Germans found, however, that they had to be on guard against dishonesty and attempts at deception by individual Russian soldiers and small units. One trick, a particular favorite, was to feign surrender, or come over to the enemy with raised hands, white flags, and all the rest. Anyone approaching in good faith would often be met by sudden surprise fire at close range. The Russian soldier, who can lie motionless for hours on end, often feigned death. An unguarded approach often cost a German his life.

Chapter 2

The Russian Command Echelons

The higher echelons of Russian command proved capable from the very beginning of the war and learned a great deal more during its course. They were flexible, full of initiative, and energetic. However, they were unable to inspire the mass of Russian soldiers. Most of the commanders had advanced in peacetime to high positions at a very early age, although there were some older men among them. All social levels were represented, from the common laborer to the university professor of Mongolian languages and cultures. Of course, merit in the Revolution played a part, but a good choice was made with respect to character, military understanding, and intelligence. Purely party generals apparently got positions carrying little more than prestige. The extraordinary industry with which the commanders went about their duties was characteristic. Every day, and far into the night, they sat together to discuss and to record in writing what they had seen and heard during the day.

During the various political purges an appreciable portion of this command strata disappeared. But it is a mistake to assume that a deterioration of the higher command levels resulted. Such great progress had been made in military education in Russia that even the higher commanders available at the beginning of the war were of a stature commensurate with their duties. In this connection it must be noted that a nation as young, energetic, and populous as Russia, was able to draw on an inexhaustible source of strength. In addition, this group of officers was held in high regard by the populace, was extolled in propaganda, and was very comfortably situated economically. Many things testified to the position which the military leaders of the Soviet Union enjoyed in the state and among the people: pictures in prewar illustrated Russian newspapers, the display in Red Army office buildings of artistically valuable portraits of senior officers and of paintings showing episodes in their military careers, and the exhibition on stairways and in government buildings of banners emblazoned with the pronouncements of high-ranking military officers.

The many developments in the sphere of strategy, which now and then gave rise to doubts about the ability of these leaders, require an examination of their background before they can be properly judged. The alleged failure during the Finnish winter campaign of 1939–40 is well known, but nevertheless the conjecture cannot be dismissed that there was some bluffing involved. The timing of the operation was correct and would produce results as soon as the will of the immeasurably superior attacker desired them.

Neither is the success of the German surprise attack at the opening of the campaign against Russia in 1941 any proof to the contrary. Along the Central Front, and also in the north, it appeared as though the actual war did not begin until the Dnepr and the Luga had been reached. One of Timoshenko's strategic war games, as well as the course of events at the beginning of the Eastern Campaign, tend to substantiate this assumption, which is supported by the following incident.

In mid-July 1941 the German LIII Infantry Corps joined the defensive battle in the Dnepr-Berezina triangle (Map 2) against Timoshenko's thrust into the flank and rear of Second Panzer Group which was advancing on Smolensk. Soon thereafter, corps found out that in February 1941, in a house on the Bobruysk-Rogachev road, Timoshenko and his field commanders had held a conference which lasted several days. Upon exhaustive search of that house a map was found indicating that German armored units were assumed to be north of Rogachev, east of the Dnepr; a pincers operation was being launched against them from the regions of Zhlobin-Rogachev and Mogilev, with Bobruysk as the objective; one very strong group of forces was to advance to the northwest via Zhlobin-Rogachev, another of like strength to the southwest via Mogilev, to cut off the German armor; the two groups were to meet at Bobruysk; the intermediary daily objectives of the two groups were indicated by semicircles; a cavalry unit of three elements, committed on the west bank of the Berezina and striking northward from the Parichi area, was to cut the Slutsk-Bobruysk road and the Minsk-Bobruysk railroad, and complete the encirclement.

Since the German armored units had already defeated the northern enemy group near Orsha, there remained only the southern arm of the pincers. With a total of 20 divisions, Timoshenko carried through this part of the operation that apparently had been planned in February. He thus came in contact with LIII Infantry Corps which had in the meantime arrived in the Dnepr-Berezina triangle by way of Bobruysk. There developed a bitter 3-week defense action between the corps—comprising three divisions—and those Russian forces which had advanced across the Dnepr near Zhlobin and Rogachev. During the course of this battle the cavalry corps indicated on the map also appeared, and for a while actually reached its objectives. The LIII Infantry Corps was cut off from the rear for about a week.

Only customs and frontier guards were encountered on the Bug; very weak enemy forces appeared after a few days; finally, the big battle in the Dnepr-Berezina triangle: thus began the Eastern Campaign. Prisoner statements furnished the Germans a very clear picture of the preliminaries, concentrations, and groupings for the battle by the Russians. Again and again reports gave the impression that large-scale enemy movements did not get under way until after the opening of the campaign, and that they took place beyond the Dnepr. From the point of view of Russian grand strategy it was undoubtedly an expedient solution. Nevertheless, the German intelligence service believed that it had identified continuous troop movements to the Russo-German frontier, supposedly involving 130 divisions, as early as spring of 1940.

The events which led up to the battle of Uman on the Southern Front offer no damaging evidence against the ability of the Russian higher command, but show only the danger and the detrimental effect of injecting politics into current military operations.

Thus the Russian high command was, for the most part, competent. Whether Timoshenko's tenacity in carrying out his plan, as mentioned above, even though the northern part of the pincers operation was knocked out, should be regarded as inflexibility, or whether Timoshenko thought that there were possibilities for a great success just the same, must be left out of the discussion. During the subsequent course of the war, however, a flexibility in strategy was apparent in the Russian high command that in the field of tactics long remained absent in the performance of the intermediate and lower command echelons. An awareness of responsibility accompanied the willingness to accept responsibility, as is shown by the following example.

After the break-through by the German LIII Infantry Corps from the southwest to Bryansk at the beginning of October 1941, the opposing Russian Fiftieth Army, commanded by General Petrov, withdrew, badly shaken, to the wooded terrain northeast of Bryansk. The Bryansk pocket was the result. During the withdrawal the commander of the Fiftieth Army had been given command of Army Group Bryansk under which, in addition to the Fiftieth Army, were placed the Thirtieth and Thirty-third Armies. The diary of Major Shabalin, a State Security (GPU) officer attached to Fiftieth Army headquarters who was killed in the Bryansk pocket, contained the following on the subject:

> I [Shabalin] congratulated General Petrov at breakfast upon his appointment as commander of Army Group Bryansk. General Petrov answered only: "So now they are going to shoot me too." I replied: "How can you talk of shooting? Your appointment as commander of the Group is an indication of confidence in your ability to get things organized again." General Petrov: "How can I get the situation back under control when I don't know where the Thirtieth and Thirty-third Armies are and what condition they are in?"

A few days later General Petrov was killed at the side of Major Shabalin, in a night attempt to break out of the Bryansk pocket.

Another example was that of Marshal Kulik who, with a serious leg injury, supported himself on a cane and led 10,000 men, whom he had reassembled after the Minsk pocket, through the swamps near Bobruysk, and gave the German troops serious difficulties. The higher Red Army commanders did not spare themselves.

The way operations were launched and carried out revealed the influence of German methods on the Soviet high command. Operations against flank and rear, large-scale envelopments, and encirclements all played a part. Other maneuvers employed were mobile defense and, finally, break-through and break-out.

Timoshenko's plan for a double envelopment and isolation of the German large armored units which had advanced across the Dnepr undoubtedly was on the grand scale. This operation was carried out energetically and efficiently. The strategic concentration, assembly, and commitment of units participating in the attack were irreproachable.

The great thrust via Yefremov at the beginning of November 1941, in which the Russians aimed at the rear of the German armored units standing in front of Tula, and which later led to the battle of annihilation southeast of Plavskoye, was likewise well planned. From a strategic and tactical point of view it had a chance of success. The weakest point of the German armored thrust had been recognized and correctly exploited.

The Russian high commands had an eye for strategically and tactically weak points of the enemy. The battle of Moscow in 1941–42 and its consequences are a good example of this as are, on a smaller scale the operations in March 1944 on the Kandalaksha front in northern Finland against XXXVI Mountain Corps. The Red Army high command can, of course, claim more and even greater successes, such as the various major offensives from 1943 to 1945.

However, the resources of their country and the large number of troop units that were available gave the Soviet command an advantage over the Germans. Equipment, training, and physical and spiritual character of their armed forces all corresponded to the conditions in the East. For this reason the Germans had to contend with a great number of difficulties which simply did not exist for the Russian high command. In addition, the low valuation placed on human life freed the Soviet high command from moral inhibitions. Whether, for example, several divisions were lost in an encirclement, or whether a reindeer division on the Murmansk front perished in a snowstorm, was of no particular importance. Not until later did the long duration of the war and the extensive losses force the Red high command to greater economy of manpower.

The flexibility demonstrated by the higher commands (army and army group) was not evident at lower levels. The lower command echelons (echelons below division level) of the Russian Army, and for the most part also the intermediate echelons (generally division level), remained for a long period inflexible and indecisive, avoiding all personal responsibility. The rigid pattern of training and a too strict discipline so narrowly confined the lower command within a framework of existing regulations that the result was lethargy Spirited application to a task, born of the decision of an individual, was a rarity. Russian elements that had broken through German lines could remain for days behind the front without recognizing their favorable position and taking advantage of it. The Russian small unit commander's fear of doing something wrong and being called to account for it was greater than the urge to take advantage of a situation.

The commanders of Russian combined arms units were often well trained along tactical lines, but to some extent they had not grasped the essence of tactical doctrines and therefore often acted according to set patterns, not according to circumstances. Also, there was the pronounced spirit of blind obedience which had perhaps carried over from their regimented civilian life into the military field. Thus, for example, toward the end of September 1941 in the area southwest of Bryansk, the same sector was attacked by various Russian battalions every day for 7 days running without any apparent reason and without success, but with severe losses. Finally, a captured battalion commander supplied the explanation. In looking through some old files, their new regimental commander had found a top-level order to the effect that continuous attacks were to be made along the entire front in order to ease the pressure on Leningrad. Since he had received a negative answer to his inquiry as to whether these attacks had already been made, he had ordered this sector attacked every day. In the meantime, however, 2 months had passed, and the pressure on Leningrad had long since been relieved.

This lethargy and reluctance to assume responsibility was a serious drawback to the Red Army, completely neutralizing a great many good points of the Russian soldier. Later on, the Soviet commanders learned a great deal along this line and became more flexible. The tactical employment of the 6th Rifle Division in the battle of annihilation southeast of Plavskoye may be termend perfect: advance, withdrawal in trucks in dangerous situations, another forward thrust, and then attack.

So far as care of the troops and internal administration were concerned, Russian unit commanders faced the same problems that confronted officers of other armies. There was great interest in hygienic measures. But also other matters, such as decorations, promotions, results of the war loan, and similar matters, kept the lower commanders very busy.

Chapter 3

The Commissar

The influence of the Communist party and of its representatives in the Army—the commissars—was tremendous. The commissar was probably the most controversial man in the Russian Army. Even in the Soviet Union opinions varied concerning his usefulness, his position, and his duties. He was the driving force of the Army, ruling with cunning and cold-bloodedness.

By means of a close-meshed network of especially chosen personalities, the commissars held the entire army machine under their control and in a tight grip. The commissars were to a preponderant degree real political fanatics. They came mostly from the working class, were almost without exception city people, brave, intelligent, and unscrupulous. But they also took care of the troops. Even though, during the course of the war, their intervention in the military conduct of the war was reduced thanks to Stalin's military instinct, their influence was not lessened.

However, it is not true that the Russian soldier fought well only because of fear of the commissars. A soldier who is motivated solely by fear can never have the qualities that the Russian soldier of this war displayed. The motive of fear may often have been the final resort in difficult situations, but basically the Russian has no less national—as distinguished from political—patriotism than the soldier of the western armies, and with it comes the same source of strength. Unceasing propaganda has burned nationalism into his soul. And however impervious he may be to foreign propaganda, he nevertheless has been unable to escape the engulfing waves of his own.

Among the troops themselves the relationship of the soldier to the commissar apparently was endurable in spite of the commissar's uncompromising strictness and severity. The higher headquarters, on the other hand, appear to have regarded him with mistrust. Testimony to that conclusion is found not only in the episode mentioned below, but also in many remarks of General Petrov, the commander of the Fiftieth Army, to Commissar Shabalin, which the latter recorded verbatim in his diary. General Petrov once ironically asked Shabalin, who was sitting next to him in the tank: "Well, how many have you shot today?" Shabalin added the note: "Such sarcasm." The commissar was thus often considered an alien element by headquarters.

The prohibition of vodka in the Russian Army which had been in effect until then, was rescinded in August 1941. A division commissar recommended that the prohibition be restored immediately and based his recommendation on an occurrence which took place at his division headquarters. The first liquor to arrive was not distributed to the troops by division headquarters, but used by the headquarters staff itself. The result was general drunkenness among the staff. The officers allegedly went out into the village street and killed geese with their pistols. When there were no more geese, an officer of division headquarters pounded the commissar on the shoulder and said, "Well, all the geese have been shot dead. Now it's your turn."

In the fighting east of Roslavl in August 1941, a Russian tank company that had been sent into action suddenly stopped on the battlefield. The leader of the tank company had received an order before going into action to refuel at a fuel depot somewhat to the rear of his bivouac area. He did not, however, want to take the trouble to go back as he thought that it would be possible to refuel farther forward at the divisional command post nearer the front. But there was no opportunity to refuel at that point. The tank company just reached the battlefield and then ground to a halt because of lack of fuel. Thereupon, the company commissar drew his pistol and shot the commanding officer on the spot.

The attitude of the common man toward the commissar was conditioned not only by fear of his power, but also by his personal exemplification of the soldier and fighter. His concern for the welfare of the troops also determined to a large extent his relationship with the men. The commissars always made much of the well-being of the troops. Commissar Shabalin, for instance, reported to higher authority the insufferable conditions on the Moscow-Orel-Bryansk railroad. While reconnoitering for new division headquarters, he immediately sent a division commissar, whom he had discovered in the rear, back to the front with the observation: "You belong with your troops; go and take care of them." There were innumerable recommendations for the improvement of conditions in the Army hospitals.

The example set by the commissars is largely responsible for the tenacious resistance of the Russian soldier, even in hopeless situations. It is not wholly true that the German commissar order, directing that upon capture commissars be turned over to the SD (Security Service) for "special treatment," that is execution, was solely responsible for inciting the commissars to bitter last-ditch resistance; the impetus much rather was fanaticism together with soldierly qualities, and probably also the feeling of responsibility for the victory of the Soviet Union. The previously mentioned occupation of the bunkers on the Bug and the continued resistance in the citadel of Brest Litvosk can be traced to the influence of the commissars.

Then, too, in innumerable other cases dogged perseverance even under hopeless conditions was to be credited to the soldierly conduct of the commissars. For instance, in September 1941, long after the castle of Posyolok Taytsy (south of Leningrad) had been taken, and strong German troop units had been drawn up in the castle park, German tanks passing near the park wall with open hatches drew single rounds of rifle fire from close range. The shots were aimed at the unprotected tank commanders who were looking out of the turrets. Not until three Germans had been killed by bullets through the head did the passing tank unit realize that the shots were coming from a narrow trench close under the park wall 10 yards away. The tanks then returned the fire, whereupon all 13 occupants of the trench met death. They were the officers of a Russian regimental headquarters, grouped about their commissar who fell with his rifle cocked and aimed.

After the German divisions broke out of the Luga bridgeheads in August 1941, the commander of a task force inspected several Russian tanks which had been knocked out 2 hours earlier near a church. A large number of men were looking on. Suddenly, the turret of one of the knocked-out tanks began to revolve and fire. The tank had to be blown up. It turned out that among the crew, which had been assumed dead, there was a commissar who had merely been unconscious. When he revived and saw the many German soldiers around him, he opened fire.

When in April 1942 the Germans took a strong position along the Osuga (southwest of Rzhev), they continued to receive rifle fire from one lone barricaded bunker. All demands for surrender were in vain. When an attempt was made to shoot through the embrasure with a rifle, the Red soldier grabbed it and fired the last three shots. Two of the bullets wounded German soldiers. The commissar, who was defending the bunker alone in the midst of his dead comrades, then shot himself with the third.

It might appear that much of the fighting spirit and concern for the welfare of the troops which the commissars displayed should have been the responsibility of the commanding officers and not of the commissars. However, it was always a question of situations in which something had to be done. The commanding officers generally did little, while the commissars acted. The passive character of the Russian officers was responsible for the fact that it was not the commander but the commissar who discovered the road to action. Therefore, the commissar was really a necessary part in the structure of the Red Army. He was a sort of front-line conscience.

It was difficult for the Russians themselves to properly judge this matter, and much more so for anyone distantly removed. Rejecting an institution which had its good points under the prevailing conditions would have been a mistake.

The commissars found special support among the women who served within the framework of the Soviet Army. Russian women served in all-female units with the so-called partisan bands, individually as gunners in the artillery, as spies dropped by parachute, as medical corps aides with the fighting troops, and in the rear in the auxiliary services. They were political fanatics, filled with hate for every opponent, cruel, and incorruptible. The women were enthusiastic Communists—and dangerous.

It was also not unusual for women to fight in the front lines. Thus, uniformed women took part in the final breakout struggle at Sevastopol in 1942; medical corps women in 1941 defended the last positions in front of Leningrad with pistols and hand grenades until they fell in the battle. In the fighting along the middle Donets in February 1943, a Russian tank was apparently rendered immobile by a direct hit. When German tanks approached, it suddenly reopened fire and attempted to break out. A second direct hit again brought it to a standstill, but in spite of its hopeless position it defended itself while a tank-killer team advanced on it. Finally it burst into flame from a demolition charge and only then did the turret hatch open. A woman in tanker uniform climbed out. She was the wife and cofighter of a tank company commander who, killed by the first hit, lay beside her in the turret. So far as Red soldiers were concerned, women in uniform were superiors or comrades to whom respect was paid.

The four elements which determine the nature of Russian warfare—the higher command, the troops, the commissar, and the Russian terrain—fitted together in such a way that their combination was responsible for good performance and great successes. The weakest elements were the intermediate and lower leaders. Their shortcomings, however, were made up for in part by appropriate action of the higher command and by the good will, the discipline, the undemanding nature, and the self-sacrificing devotion to duty of the enlisted men under the influence of energetic commissars who were filled with a belief in the essential necessity of victory. The Russian soldier thereby became an instrument which provided his leaders with the sort of fighter needed for the operations.

Chapter 4

The Combat Arms

I. Infantry

The picture of the Russian soldier presented up to this point applies primarily to the infantry. This branch supplied the bulk of the fighting men in active combat, and most clearly exemplified the peculiar characteristics of the Russian. The Red Army infantry was for a long period of time the mainstay of the Russian fighting machine, and in the Arctic this was true for the entire war period. As is apparent from the previously described characteristics of the Russian soldier, the Soviet infantry was willing, undemanding, suitably trained and equipped, and, above all, brave and endowed with a self-sacrificing devotion to duty. The communist philosophy appeared to have become firmly rooted among the great mass of the younger people and to have made them loyal soldiers, differing much in their perseverance and performance from those of World War I. Subject to rapidly changing moods, the Russian infantryman was jovial one moment, cruel the next.

The Russian infantryman was a member of the herd, preferring to fight in concert with others rather than to be left to his own devices. In the attack, this characteristic was evidenced in the massed lines, sometimes almost packs; in the defense it was shown by the stubbornly resisting bunker complements. Here, there was no individual action for one's personal advantage. The soldiers aided each other and even displayed an interest in their comrades' family affairs.

As has already been pointed out the Russian soldier was virtually immune to seasonal and terrain difficulties. Further, he was almost complete master of the terrain. There appeared to be no terrain obstacles for the infantryman. He was as much at home in dense forest as in swamp or trackless steppe. Difficult terrain features stopped him only for a limited time. Even the broad Russian streams were crossed quickly with the help of the most primitive expedients. The German could never assume that the Russian would be held back by terrain normally considered impassable. It was in just such places that his appearance, and frequently his attack, had to be expected. The Red infantryman could, if he chose, completely overcome terrain obstacles in a very short time. Miles of corduroy road were laid through marshy terrain in a few days; paths were tramped through forest covered with deep snow. Ten men abreast, with arms interlocked, and in ranks 100 deep, prepared these lanes in 15-minute reliefs of 1,000 men each. Teams of innumerable infantrymen moved guns and heavy weapons wherever they were needed. During the winter, snow caves which could be heated were constructed to furnish night shelter for men and horses. The Russian matériel was useful in this respect: motorization reduced to an absolute minimum; the lightest vehicles; tough horses that required little care; suitable uniforms; and, finally, again the human mass which moved all loads and performed all required tasks like a machine.

A singular kinship to nature makes the Russian infantryman an ideal fighter in forests, barren country, and swamps. In the Arctic, small units of several men stayed for weeks in the desolate area in the German rear, accompanied at times by female radio operators who were treated with particular respect.

Wounds were endured patiently and without complaint. Frostbite was a punishable offense because it was avoidable. Recovery, even from serious injuries, was rapid. Many an injury that would have been fatal to a Central European was endured and overcome.

The best weapon of the Russian infantryman was the machine pistol. It was easily handled, equal to Russian winter conditions, and one which the Germans also regarded highly. This weapon was slung around the neck and carried in front on the chest, ready for immediate action. The mortar also proved highly valuable as the ideal weapon for terrain conditions where artillery support was impossible. At the beginning of the Eastern Campaign, Russian infantry far surpassed the German in mortar equipment and its use.

The same was true for the Russian antitank gun, which at the beginning of the campaign considerably surpassed the antitank gun of the German infantry divisions in efficiency, and therefore was readily put to use whenever captured. The antitank gun was an auxiliary weapon from which the Russian soldier never separated. Wherever the Russian infantryman was, antitank defense could be expected by his enemy. At times it appeared to the Germans that each Russian infantryman had an antitank gun or antitank rifle, just as infantrymen of other armies had ordinary rifles. The Russian moved his antitank defense everywhere with great skill. It was to be found even where no German tank attacks might be expected. Emplacements were set up within a few minutes. If the small gun, always excellently camouflaged, was not needed for antitank defense, its flat trajectory and great accuracy were put to good use in infantry combat. The Germans had a rule to cope with this: Engage Russian infantry immediately following their appearance, for shortly thereafter not only the soldier but also his antitank defense will have disappeared into the ground, and every countermeasure will be twice as costly.

In World War II the Russian infantryman had a noteworthy negative characteristic: He was not inquisitive. His reconnaissance often was extremely poor. Combat patrols were for him the means of gathering information about the enemy only when he thought it necessary. Although the Russian proved himself an excellent scout, he made too little use of his abilities in this field. The higher Russian command was always well informed on the German situation by means of radio monitoring, interrogation of prisoners, captured documents, and other means. But the intermediate and lower commanders apparently were only slightly interested in their opponents. Here again the cause lay in the lack of self-reliance and in the individual Russian infantryman's inability to assemble into a useful report the observations made while on patrol.

The clothing and equipment of the Russian infantryman suited his summer as well as winter requirements. The Germans were amazed at how well the Siberian infantry was clothed in the winter of 1941–42. As might be expected, the fighters in the Arctic were likewise suitably clothed. The Russian infantryman was inferior to the German and to the Finn only in skiing. Of course, attempts were made to correct this deficiency through intensified training, but all efforts were doomed to failure since there was never more than one pair of skiis available for several men.

Equipment carried by all Germans was often discarded by the Russian infantryman as nonessential. Gas masks were commonly stored in division depots; steel helmets were rarely worn in the arctic wilderness.

II. Artillery

The efficiency of the Russian artillery varied greatly during the various stages of the war. In the beginning it was unable to achieve an effective concentration of fire, and furthermore was unenthusiastic about firing on targets in the depth of the battle position even when there was excellent observation. The Rogachev water tower, for example, and the railroad control towers as well as the high railroad embankment at Zhlobin, all of which were in Russian hands during the battle in the Dnepr-Berezina triangle, commanded a view over the entire area for many miles; nevertheless, they were not used for directing fire on the very important targets behind the German lines. On the Kandalaksha front, continuous German supply transport operations at the Karhu railroad station took place within sight of Russian observation posts. These operations were never taken under fire by Russian artillery. On the other hand, the Russian artillery liked to distribute its fire over the front lines, and occasionally shelled a road intersection located not too far from the front.

During the course of the war the artillery also developed to a high degree the use of mass as a particularly characteristic procedure. Infantry attacks without artillery preparation were rare. Short preparatory concentrations lasting only a few minutes, frequently employed by the Germans to preserve the elements of surprise, seemed insufficient to the Russians. Thus, counting on the destructive effect of massed fire, they consciously accepted the fact that the Germans would recognize their intentions of attacking. Russian artillery fire often had no primary target, but covered the entire area with the same intensity. The Russian artillery was most vulnerable to counterbattery fire. It ceased firing or changed position after only a few rounds from the German guns. The rigidity of the fire plan, and a certain immobility of the Soviet artillery—at least during the first years of the war—was pronounced. Only in rare cases was the artillery successful in promptly following the infantry. Most of the time the artillery was unable to follow up; it remained stuck in the old positions, leaving the infantry without fire support. This practice frequently took the momentum out of the Russian attacks.

Attack tactics of Russian artillery improved constantly during the war. Eventually, however, their tactics resolved into an ever-repeated, set scheme. Heavy preparatory fire, laid down broad and deep and lasting from one to two hours, was the initial phase; it rapidly mounted to murderous intensity. Once an attack was about to get under way, the Russians would suddenly lift their fire from very narrow lanes (about 80 to 100 yards wide), along which the infantry was to advance. At all other points the fire continued with undiminished fury. Only the most careful German observation allowed recognition of those lanes. This method gave the impression that artillery preparation was still continuing in full force, though in reality the infantry attack had already begun. Here again, one notes the same concept: human lives meant nothing at all. If defensive fire forced the Russian infantry out of their narrow lanes, or if their own artillery was unable to maintain the lanes accurately—*Nichevo!*—those were operating expenses.

However, despite many shortcomings, the Russian artillery was a very good and extremely dangerous arm. Its fire was effective, rapid, and accurate. Particularly during the large-scale attacks in the summer of 1944 it became apparent that the Russians had learned well how to mass and employ large artillery units. Establishment of a definite point of main effort and the use of superior masses of artillery crushed the thin lines of German opposition at many places at the Eastern Front before the actual attack had begun. This successful procedure of establishing definite points of main effort will be used by the Russians in the future whenever they have the masses of artillery and ammunition required.

III. Armored Forces

The heart of the Russian armored force was the well-known T34 tank. Because of its wide tracks, its powerful engine, and its low silhouette, the performance of the T34 in Russian terrain was frequently superior to that of the Germany tanks, particularly with respect to cross-country mobility. To the surprise of all the German experts, the T34 easily negotiated terrain theoretically secure against mechanized attack. The caliber of its guns was too small, however, and forced the Russians to produce several new types of tanks (KV1 and Stalin) which, like the German models, became successively heavier. Despite all improvements the new tanks remained on the whole inferior to the T34. The Russians recognized this fact, and continued to mass produce the T34 until the end of the war.

Not until late did the Russians decide to launch concerted attacks by large tank forces. During the first years of the war, Russian tanks generally were used for local infantry support. Soviet tank attacks as such took place only after a sufficient number of the vehicles had become available. Here, too, the Russians adhered to their usual habit of employing great masses of men and machines.

Tank attacks generally were not conducted at a fast enough pace. Frequently they were not well enough adapted to the nature of the terrain. Those facts the Germans noted time and again through the entire war.

The training of the individual tank driver was inadequate; the training period apparently was too short, and losses in experienced drivers were too high. The Russian avoided driving his tank through hollows or along reverse slopes, preferring to choose a route along the crests which would give fewer driving difficulties. This practice remained unchanged even in the face of unusually high tank losses. Thus the Germans were in most cases able to bring the Russian tanks under fire at long range, and to inflict losses even before the battle had begun. Slow and uncertain driving and numerous firing halts made the Russian tanks good targets. Premature firing on the Russian tanks, though wrong in principle, was always the German solution in those instances. If the German defense was ready and adequate, the swarms of Russian tanks began to thin out very quickly in most cases. This fault in Russian tank tactics can be corrected only by peacetime training, but it can hardly be totally eliminated.

On the whole, the Russian armored force was not as good as the Russian artillery. Limited flexibility, and the inability of the subordinate commanders to exploit favorable situations rapidly and adroitly, were evident and frequently prevented the Russians from achieving successes almost within their grasp. Toward the end of the war, however, the inadequate facilities of the Germans were no longer able to stand up against the masses of equipment of the Reds.

IV. Horse Cavalry

In the campaign the Russian cavalry, despite many changes in tactics and equipment, achieved a significance reminiscent of old times. In the German Army, all cavalry except one division had been replaced by panzer units. The Russians followed another course. The German LIII Infantry Corps quite often encountered Russian cavalry divisions, and once a cavalry corps comprising three cavalry elements—always in situations in which cavalry was a suitable arm for the purpose.

In the battle of the Dnepr-Berezina triangle, a cavalry corps comprising three elements appeared west of the Berezina near Bobruysk, out of the Pripyat Marshes, in the rear of the German corps which was engaged in hard fighting. This cavalry force cut the Slutsk-Bobruysk highway and the Minsk-Bobruysk railroad, and thereby isolated the corps for a week from its supply and contact with the rear. Bobruysk itself, together with the bridges there, was seriously threatened. Only by prompt emergency measures were the Germans able to ease the pressure. During that period the corps ammunition supply dropped to twenty rounds per 105-mm. gun, and that at a time when Timoshenko's major offensive had reached its peak. Forces other than cavalry would have been unable to conduct such a raid.

In the advance out of the Bryansk pocket northward past Plavskoye to the upper Don in October–November 1941, the 112th Infantry Division of LIII Infantry Corps was met by a Russian cavalry division just as it reached the Orel-Tula road. There ensued some very unpleasant delays. The cavalry division accomplished its mission by occupying every town along the route of advance, withdrawing from each one as soon as the division advanced to attack. Thus the 112th Division was kept occupied by constant small-scale warfare.

During the battle of annihilation southeast of Plavskoye, cavalry divisions appeared on both wings of the Russian attack front. The cavalry division just mentioned was on the northern wing. The Russian command had diverted it to the battlefield as soon as the 112th Infantry Division turned in that direction. The cavalry division, with its greater speed, reached the Plavskoye area ahead of the 112th Division and blocked its advance until a Russian rifle division arrived at the front. The cavalry division mentioned earlier was on the southern wing. That division was to reach the Orel-Tula highway in a forced march so as to cut it near Chern—under prevailing circumstances, a rewarding and practicable task for a cavalry division. The corps' further advance to the upper Don was constantly accompanied and watched by a cavalry division 5 to 10 miles off its right flank. The cavalry division would have been able to intervene at once, in any manner whatsoever, if the Russian higher command had had any occasion for ordering it to do so.

The above examples are drawn from situations in which the missions in question could have been executed only by cavalry. The German armored vehicles were out of action in October during the muddy period. Only cavalry could operate through the Pripyat Marshes.

Under conditions as characterized in Central Russia by great forest and swamp areas, muddy periods, and deep snow, cavalry is a usable arm. Where the German motor failed, the Russian horse's legs continued to move. The tactical employment of the cavalry forces was, however, not always suited to the situation and sometimes was even awkward. Leadership and training in the Russian cavalry were not up to the World War I standard.

Chapter 5

Russian Battle Techniques

Just as the Russian soldier had his own peculiarities in his internal make-up, so he had them in his combat methods. The most common Russian form of combat was the use of mass. Human mass and mass of matériel were generally used unintelligently and without variation, but under the conditions, they were always effective. Both had to be available before they could be used so lavishly and were therefore dependent upon limitless Russian supplies. The Russian disdain for life—always present, but infinitely heightened by communism—favored this practice. A Russian attack which had been twice repulsed with unheard-of losses would be repeated a third and a fourth time at the same place and in the same fashion. Unimpressed by previous failures and losses, new waves always came on. An unusual inflexibility of mind and unimaginative obstinacy lay in this use of masses, and was dearly paid for. It is not possible to estimate Russian casualties in World War II with any degree of accuracy; there will always be a potential error of many hundred thousands. This inflexible method of warfare, with the objective of accomplishing everything through the use of human masses, is the most inhuman and costly possible.

Characteristic of the disdain for human life was the complete elimination of military funeral rites. There was no such thing as a funeral ceremony for the ordinary citizen in communist Russia. It ran counter to the antireligious philosophy and the mass sacrifice of human beings. There were occasional burial mounds in which a thick peg, painted red, had been driven into the ground at the head end, inscribed with the heroic deeds of some commissar killed in action. The bulk of the dead millions were unceremoniously plowed under. Thus, all outward indications of the number of dead were obliterated.

Not until 1944 did the Russians start using their men more sparingly. Only after there were no new millions available did there appear in orders the first reference to consideration in the use of men and admonitions to avoid losses. Nevertheless, a thinning of the attack waves was virtually unknown to the Russians until almost the last days of the war. The herd instinct and lack of self-reliance on the part of the subordinate commanders time and again misled them into the concentrated employment of troops. It was not particularly difficult to crush the attack of the human mass so long as the opposition possessed a mass of material in trained hands.

In the winter of 1941, the Russians cleared a German mine field south of Leningrad by chasing over it tightly closed columns of unarmed Russian soldiers shoulder to shoulder. Within a few minutes, they became victims of the mines and defensive fire.

The inflexibility of Russian methods of warfare was evidenced repeatedly. Only the top Russian command during the last years of the war was an exception. This inflexibility manifested itself as high as army level; in divisions, regiments, and companies it was unquestionably the retarding factor in the way the Russians fought. A division boundary was a sacred wall, and a neighbor's interest halted at his side of that wall. The senseless repetition of attacks, the rigidity of artillery fire, the plotting of lanes of attack and movement without regard to terrain, were all additional symptoms of inflexibility. The leaders displayed a certain flexibility in their frequent shifting of units in the front lines. These units disappeared unnoticed overnight and reappeared several days later in another sector of the front. It was, however, no feat to relieve and exchange troops so long as one had the reserves that the Russians had; on the other hand, it was all the more difficult if one had none, as was the case with the Germans. The Russians have this method to thank for the fact that only a few of their units became thoroughly depleted during the course of prolonged battles; on the German side such a depletion became inevitable over the years of continuous service of units. None of the Soviet units that were thus shifted from sector to sector ever disappeared. It could be safely assumed that the unit withdrawn would be recommitted within the same army sector. The organization of an army and the assignment of divisions to it also remained fixed. It seldom happened that divisions were interchanged between the armies.

Though generally no master of improvisation, the Russian command nevertheless knew how to bring battered infantry units up to strength and how to constitute new units as replacements for destroyed ones. This procedure frequently was accomplished with startling speed, but it soon turned out to be a game played with human lives. For example, the inhabitants of a threatened city, or perhaps the entire male population of areas which the Germans had recently evacuated, were gathered up quickly by means of excellent organization. Regardless of age, nationality, deferred status, or fitness, they were used to fill out these units. With no training at all, or at most only a few days of it, and often without weapons and uniforms, these "soldiers" were thrown into battle. They were supposed to learn in combat all that was necessary, and to acquire their weapons from their dead comrades. The Russians themselves were aware of the fact that these men were no soldiers, but they filled gaps and supplemented the sinking numbers of the human mass. During the fighting in a bridgehead southeast of Krememchug in September 1943, the Russians at nighttime used to drive ahead of their armed soldiers large numbers of civilians whom they had gathered up, so that the German infantry might expend its scant supply of ammunition.

The Russians repeated the same tactics again and again: employment of masses, and narrow division sectors held by large complements replenished time after time. Therefore also the mass attacks. In the twinkling of an eye the terrain in front of the German line teemed with Russian soldiers. They seemed to grow out of the earth, and nothing would stop their advance for a while. Gaps closed automatically, and the mass surged on until the supply of men was used up and the wave, substantially thinned, receded again. The Germans often witnessed this typical picture of a Russian attack. It is impressive and astounding, on the other hand, how frequently the mass failed to recede, but rolled on and on, nothing able to stop it. Repulsing such an attack certainly depended on the strength of the forces and means for defense; primarily, however, it was a question of nerves. Only seasoned soldiers mastered the fear which instinctively gripped everyone upon the onslaught of such masses. Only the true soldier, the experienced individual fighter, could in the long run stand up under the strain; only a multitude of them could stop those masses.

Another specifically Russian battle technique was infiltration. It was a practice which especially suited the Russian, and of which he was a master. Despite closest observation of the avenues of approach, the Russian was suddenly there; no one knew where he had come from, nor how long he had already been there. Wherever the terrain was considered impassable, but was still kept under close observation to be doubly safe—just there the Russian infiltrated. He was suddenly there in substantial numbers and had already vanished into the earth. Nobody had seen a thing. Because of the drawn-out German defense fronts, it was no particular art to steal between the widely separated strong points, but it was always a surprise when, despite all watchfulness during the night, the Germans found the next morning that strong Russian units fully equipped with weapons and ammunition had assembled and dug in far behind the front. These operations were executed with unbelievable skill, completely noiseless and almost always without a struggle. It was a very profitable technique which succeeded in hundreds of cases and gained the Russians great advantages. There was only one method of countering it: extreme watchfulness, and heavily occupied, deep positions, secured throughout at all times.

Of the same nature was the Russian's constant effort to establish bridgeheads (or advance covering positions, which are here included in the same category). These bridgeheads often served to harass the German and to sap his strength, and often were used as a base for

Russian attacks. They were established by infiltration or by attack, and were a dangerous Russian tactic. It always proved wrong and absolutely fatal to do nothing about such bridgeheads or to postpone their elimination, no matter what the reasoning. It was certain that Russian bridgeheads which had existed only 24 hours would during that time have grown into a serious menace. Though only one Russian company might have occupied the newly formed bridgehead in the evening, by the next morning it was sure to have turned into an almost invincible fortress held by at least a regiment and bristling with heavy weapons. No matter how heavy and accurate the German fire, the flow of men into the bridgehead continued. Regardless of all countermeasures the bridgehead continued to swell until it ran over. Only by using very strong forces and planned attack could it still be contained or eliminated, provided one was lucky and not afraid of heavy losses. Therefore, the warning against these bridgeheads can never be stressed sufficiently. There was only one way of fighting them, and that had to be made a rule: Every Russian bridgehead in the process of being formed, and every advanced position, no matter how small, must be attacked *immediately*, while it is still undeveloped, and eliminated. If the Germans waited even a few hours, they were in most cases too late, and on the next day success was more remote than ever. Even if there were but one platoon of infantry and one tank available, they had to be committed at once; the Russian was not yet dug in, he was still visible and within reach, he still had no organized defensive fire and no heavy weapons in the new position. A little later he would have all of these. Every delay brought disaster; only an immediate counterattack led to success.

The Russian's kinship with nature expressed itself also in his preference for night operations. Shifts from one position to another, and concentrations, executed rapidly and silently in darkness, were more frequent than night attacks, which themselves were no rarity. However, the Russian generally avoided large-scale night attacks, since it appeared to him that the individual soldiers and subordinate commanders lacked the necessary self-reliance. A major offensive operation was conducted at night only when a victory definitely expected by the highest command had not been achieved in the course of daylight attacks. Such large-scale night attacks were mostly acts of desperation that failed to produce success. On the other hand, night attacks with a limited objective were a method which the Russians liked to employ in order to gain advantages for the next day, or to recapture lost ground. These night attacks were mostly infantry battles only, without use of artillery. The Germans found that it paid to be prepared for a Russian attempt to regain in the night a position that had been lost during the day.

The Russian's excellent utilization of railroads should be noted, even though it does not actually fall into the category of combat. The Russians accomplished strategic concentrations or shifts of large units, one or more armies, in an unbelievably short time with the help of the railroads. The imperfect rail net of European Russia did not often present opportunity for such measures, but whenever a chance offered itself, it was thoroughly utilized. Every expedient was used even though it bore the unmistakable mark of improvisation: a continuous succession of trains without any technical railroad safety measures; trains traveling one within sight of the next, or on two tracks in the same direction; reckless overcrowding and overloading of trains. The number of trains necessary per unit was less here than for any other army. The Russian neither had nor needed baggage or accessories, clerk's bulky lockers, surplus vehicles, or large food supplies. He could be packed into railroad cars like a sardine in a can. The wider gauge and the corresponding greater width of the Russian railroad car was significant in this respect. In any event, by using this type of railroad transport the Russians succeeded in surprising the German command, since such rapidity in large-scale troop movements by railroad contradicted all experience.

Fighting the Russian on his own ground necessitated orientation in a new type of combat. Fighting had to be primitive and unscrupulous, but rapid and flexible. The German could never afford to hesitate even in the face of the most surprising occurrences and unexpected turns of events. Russian reactions to the tactical rules of warfare, and to tricks of the trade such as envelopment, flanking threat, feint, surprise, etc., could never be determined beforehand. One time the Russian would react according to the rules and give up sooner than anticipated; on another occasion all efforts were to no avail, and neither encirclement nor flank attack fazed him. The struggle with the Russians was thereby placed on an entirely new basis; predictive calculation was useless, and every action was full of suspense and surprise. At times, positions which were tactically decisive according to normal concepts would be evacuated by the Russians without delay. At other times, individual terrain features of no evident importance would be held even in the face of serious losses. In those instances the Russian often relied only on his intuition, which he substituted for his practical schooling. It must be conceded that this tactical instinct often stood the Russian in better stead than the theories of many academies could have done. The soundness of many an action which at first seemed surprising, often had to be recognized at a later time.

There was just one tactical misconception which the Russians could not relinquish despite all losses: the belief that an elevation was in every case the only terrain feature of value. They tried for every elevation, fought for it with perseverance and strong forces, without estimating whether it actually dominated the terrain or whether the situation demanded its possession. If the elevation was really important, the Germans thought it well to occupy it before the Russians; otherwise, it was to be expected that the elevation could be taken only with heavy losses. In more cases than had earlier been assumed, however, the possession of level ground was sufficient; the value of the elevation was frequently overestimated.

Thus the Russian soldier voided a number of tactical principles. Still others were altered because of the nature of the country itself, a factor discussed in the following chapter.

Chapter 6

Russia as a Theater of Operations

In the characteristics of Russian terrain, the Germany Army was faced with an inordinate number of new experiences which made necessary the formulation of new doctrines of combat. The Germans had to become thoroughly familiar with Russian land and climate—new enemy weapons to them—in order to deal with them, or at least, to neutralize their restrictive effects. Their qualities altered tactical procedures formerly in common use, and forced the adoption of measures necessary for military operations in that particular area and climate.

In many instances, climate and soil conditions in the U. S. S. R. would for extended periods of time void all rules governing military movement. The German Army was not modern enough to overcome these difficulties, and therefore was forced to fight in winter and during muddy periods as best it could without the assistance of operative motor vehicles. This retrogression of about a century was a problem which could be dealt with only by improvisation.

The great forest and swamp areas of Arctic, Northern, and Central Russia often forced all movement into the narrow lanes of a few sand and swamp roads, and made it extremely difficult and time-consuming. Development of the columns was generally out of the question. The execution of any and all measures required a multiple of the usual amount of time, and the advantage of motorization was almost negligible.

Sand, mud, and snow impeded the advance of all types of troops, and put great strain on motors. Lined up and jammed along one road, the troops slowly and painfully inched forward—a long snake crawling slowly over the difficult ground. The halt of one part of the snake stopped all parts behind it, pressing them still closer together. Weak bridges barely supported the infantry vehicles. Extensive bridge and road repairs were necessary, and many defiles had to be traversed. Previous experiences and doctrines were useless under these conditions. Time calculations, the most important factor in operations, had to be based on entirely new concepts. Frequently the time required for a movement could not be estimated even approximately. It continued to run far in excess of the expected maximum. Remedies were difficult and could be achieved only by flexibility and practice.

In by far the largest part of Southern Russia, and in most of the swampless areas of Northern and Central Russia, travel in dry weather was considerably better, and the terrain, too, was passable for vehicles of all types almost everywhere. During that time, operations on even the largest scale were possible everywhere with the exception of the swamp areas.

The principle of establishing zones of movement rather than routes of advance proved satisfactory for the Germans as well as for the Russians; the same was true of the separation of zones of movement for motorized and nonmotorized units. Here, the German motorized troops had to be allotted the zones with the best roads. Engineers were always placed far up in the column so that they might quickly reinforce bridges and eliminate terrain obstacles. Only strict regulation of traffic and relentlessly severe discipline guaranteed the best possible progress. Every halt for rest had to be closely timed, passing procedure strictly regulated, and priority definitely determined. Each driver who pressed forward in unguarded moments became a traffic violator. For that reason, trained troops were needed for traffic control, and had to be used, just as every tactical unit, according to the point of main effort. Likewise, all staffs and higher headquarters had to abide by traffic discipline since exceptions only caused confusion. In particularly difficult large-scale movements, special purpose staffs under the command of senior officers were employed.

The rate of progress of motorized troops could be roughly estimated in advance only if they were traveling on hard-surfaced roads. A rule of thumb was that on average dry roadways, motorized troops made from 5 to 6 miles per hour, and foot troops from 1 to less than 3 miles per hour. Terrain and road conditions in the East required, according to German experience, three to four times the amount of motor fuel needed under normal conditions in Europe.

The assumption that long columns can easily be stopped by the enemy appears justified. However, it developed that the Russian overestimated the constraining effect of channeled movements. In almost every case he blocked only the roads and bridges with task forces deployed without breadth or depth. It was, therefore, not too difficult to knock out such enemy blocking attempts by attacks against the flanks and rear of the enemy groups. This maneuver required two conditions for success: reconnaissance far to the front, and incorporation of combat forces capable of cross-country movement at the head of the march groups. No matter how easy it was for vehicles with cross-country mobility to cross great, open stretches in any direction during dry seasons, on long marches they were confined mostly to the existing roads, since the Russian terrain always confronted them with terrain obstacles which they were unable to overcome (swamps, watercourses, ravines, steep rises, etc.). Furthermore, it required considerably more time and fuel to move cross-country.

In drawing any conclusions from the battles of the German armies in the East it is to be noted that the German armed forces went unprepared into the Russian winter and into the muddy periods, and for that reason they faced numerous difficulties and emergencies which could have been avoided by conscientious preparation.

PART THREE

PECULIARITIES OF RUSSIAN TACTICS

Chapter 7

General

An understanding of Russian tactics demands a clear conception of the premises on which they are based. Russian tactics have their roots in the Russian himself and in his social order; in the Russian land mass with its climate and soil conditions, its expanse, and its varied topography. These points will be described in detail in the following pages, and illustrated by numerous examples. The examples should permit insight into the practical effect of these characteristics under various conditions, and should illustrate how the German command coped with them.

While the bulk of the Red Army consisted of infantry divisions which were combined into corps and armies and supplemented by numerous GHQ troops (artillery, antitank, and engineer brigades, as well as tank and self-propelled artillery elements), there were also the large, highly mobile tank and mechanized corps—the strategic force—which in most instances were combined into tank armies. Tank corps and mechanized corps were seldom placed under infantry armies. This method of organization reveals the effort of the Russian high command to assure strategic mobility. That mobility was demonstrated in the sweeping employment of these large, fast forces. As soon as the battle fronts became static (position warfare), the Russian command withdrew its tank and mechanized forces, and did not commit them again until the infantry had breached the enemy defensive system. This mode of employment coincided with the concepts expressed in the Russian field service regulations. In fighting for a system of defensive positions, GHQ tank and self-propelled artillery regiments (that is, not the organic forces of the mobile corps) supported the infantry divisions in great numbers.

Because of the long duration of the war it was possible to observe the combat methods of the Russian in all phases. During the first winter in Russia it was evident that the German lower command would have to find new methods of fighting if it wanted to be successful. The extensive area, the mass employment of men, the peculiar characteristics of Russian fighting methods, and the different climatic conditions forced the Germans to develop a new line of tactical doctrine. In logical sequence, this doctrine was derived from close observation of all Russian habits, Russian reaction to German fighting methods, and thorough exploitation of Russian weaknesses. Only a better-led soldier, and one superior in fighting qualities, had any prospect of success. In order to make details about the characteristics of the Russian available to all concerned, it was exceedingly important to compile the information and to disseminate it, particularly to the replacement training units and to the schools. There had to be a constant flow of instructors with combat experience to these units in order to train replacements and new recruits according to the latest experiences.

These measures proved invaluable to the Germans. Divisions that had been newly formed, or transferred from other theaters, and sent to the East without enough officers and noncommissioned officers with prior experience in the East, failed in the beginning almost without exception and suffered heavy losses. On the other hand, even remnants of divisions which had long been going through the hard school of the East, made outstanding accomplishments, both in Russia and on all other fronts.

Chapter 8

Reconnaissance and Security

Russian march reconnaissance and security generally corresponded to German concepts. Activity was not especially intense, but very skillfully adapted to the objective and to local conditions (terrain, climate, transportation routes, and weather). In his march reconnaissance the Russian was very hesitant, groping, and cautious, and allowed himself much time. If he found no enemy resistance, he frequently plunged ahead into the unknown without further reconnaissance.

The engagement at Maloryta at the end of June 1941 grew out of the thrust of a Russian rifle division east of the Bug near Wlodawa into the flank of German armored units which were rolling from Brest Litovsk through Kobryn in the direction of Slutsk. The Russian division had sent ahead armored reconnaissance cars. These made their appearance about half a day in advance of the division itself, and encountered the German 255th Infantry Division which was advancing from Wlodawa on Maloryta. Although the terrain (Pripyat Marshes) was extremely unfavorable for reconnaissance by armored cars, they advanced until two cars had been knocked out. The others withdrew. Instead, there appeared after a few hours infantry security detachments, already deployed as a thin screen of skirmishers. Several miles behind followed the rifle division split into battle teams, which were deployed over a front of about 6 miles, with large gaps between the teams.

The rifle division had undoubtedly anticipated an encounter with the Germans in the vicinity of Maloryta, and had sent out motorized reconnaissance to obtain the information needed for deployment of the division. The infantry screens following later served as local security for the battle teams.

In an action southeast of Plavskoye in November 1941 the Russians intended to thrust from Yefremov along the main road to the northwest in the direction of Tula, into the rear of German armored units which were before Tula. (Map 5) When the leading elements of the German 167th Infantry Division met them near Teploye, the Russians were as surprised as the Germans. It is certain that they had not counted on meeting the enemy there. On this basis, their advance can be taken as typical of a Russian march against the enemy with only one usable road available, when direct contact with the enemy is not yet expected. Again, several armored reconnaissance cars appeared at first, but they seemed to have only local reconnaissance missions. They moved ahead by bounds. Behind them, following German procedure, came infantry security detachments. Only then followed the foremost division (6th Rifle). This division possessed numerous trucks which, however, were not used until later during local shifts on the battlefield. They were not intended for the strategic forward displacement of the division at this time. On the road behind the 6th Rifle Division, further infantry divisions with fewer trucks followed. Several days later, when the battle was already in progress, a cavalry division which had come up on a bad side road running to the west appeared. A tank brigade which had been standing ready behind the long column of rifle divisions was thrown forward in a single movement only when the battle had reached its climax.

It seemed as if in an advance against the enemy, the chain of rifle divisions following one another was not to be disturbed. Accordingly, though it would have been possible to move a single division forward, this was not done. Perhaps this maneuver was also made to prevent the enemy from recognizing the movement prematurely. Had the cavalry division advanced simultaneously with the rifle divisions, the Russians would not have been able to surprise the Germans.

The following example from the Kandalaksha front shows the use of stationary security detachments for the protection of a flanking movement. About 1 mile in front of German strong points in the no man's land of primeval forest north of the Arctic Circle there was a small range of hills, the Ahvenvaara. It was unoccupied most of the time, but now and then either side would occupy it temporarily as an anchor point for operations. One day in March 1944 this range of hills was again occupied by weak Russian forces. Assuming that the Russians would, as they had for years, leave again after a short time, the Germans took no action. In a few days, however, a whole battalion had infiltrated, and it was too late to dislodge them. In the same way, the Russians occupied an elevation farther to the west, where they drove off weak German security detachments and immediately set up a strong point. Under the protection of these two points, the Russians moved their attack troops into the northern front. This was an instance of planned, far-sighted preparation for moving attack troops into the jump-off position for a subsequent attack based on a stabilized front. The Russians were completely successful in their security measures for those movements.

The preceding three examples reveal that the Russian demonstrated great adaptability in march reconnaissance and security, and that he knew how to take proper action under diverse circumstances

In mobile warfare the use of fast-working motorized reconnaissance can be observed on approaching an expected enemy, and as its result the developed advance of battle groups under cover of local infantry security detachments. In advancing upon an enemy with whom an engagement is not anticipated in the near future, the Russian normally employs an almost schematic local march-reconnaissance pattern: armored reconnaissance cars moving ahead in rapid bounds, and behind them normal infantry march security. There also seems to be a definite effort to hold the forces together and to prevent ventures on the part of individual elements. Further, in a kind of position warfare—even in primeval forest wilderness—first there is the methodical establishment of firm strong points, under the protection of which the march movements of the units destined to attack are subsequently to be executed. Surprise attacks into the Russian march movement—in the example presented, such surprise attacks would have meant a flanking march of almost 40 miles—are thereby to be eliminated.

On stabilized fronts, the Russian conducted his reconnaissance with patrols, or by reconnaissance in force. He was exceedingly adept at combat reconnaissance in offensive as well as in defensive operations. He knew how to adapt his reconnaissance patrols to the terrain and how to employ them in great numbers. Seldom could any conclusion as to the intentions of the Russian enemy be drawn from his reconnaissance-patrol activity.

For reconnaissance in force, the Russian employed forces at least in company strength, but frequently also in battalion and regimental strength. They were supported in their operations by concentrated artillery fire, and often by accompanying tanks as well. The main objective of such attacks was the taking of prisoners, and sometimes the capture of an important terrain feature.

Another frequently employed Russian method of reconnaissance was the ambush of enemy reconnaissance patrols. The Russian was a master at this procedure. Well camouflaged, he could lie in wait for hours until the opportunity presented itself. He also knew how to sneak up and surprise isolated outguards. The employment of dogs trained in tracking ambush patrols proved a satisfactory German countermeasure.

The Russians also made considerable use of the civilian population for intelligence missions, using them to obtain the desired reports on the enemy situation. A favorite practice was the employment of boys 8 to 14 years old. They were first trained for this work and then allowed to infiltrate at suitable front sectors. Before the offensive in July 1943, for example, more than a dozen such children were picked up in the Byelgorod sector alone. They gave detailed reports on the kind of training they received and on their *modus operandi.* The training of these children had been supervised by Russian officers. It had lasted 4 weeks, and there had been 60 participants. The youths came from communities near the front on both sides of the battle lines and therefore were thoroughly familiar with the locale. Many were staying with relatives or acquaintances in German-occupied localities, and were therefore not easy to discover and apprehend. Their talent for observation and their skill at spying were remarkable. For this reason, civilians in localities near the front (4 to 6 miles behind the front line) had to be evacuated not only because of the danger from enemy fire, but also as a preventive measure against espionage.

Chapter 9

Offense

The national characteristics, which have already been described, and the peculiarities of the country were the principle factors determining the Russian conduct of battle.

I. Winter: the Preferred Season

The Russians preferred to carry out their major offensives in winter because their troops were accustomed to that season and very well equipped and trained for it. The Russians were superior to all peoples of Central and Western Europe in enduring rigors of weather and climate. Casualties from the cold were an exception in the Red Army. Soldiers with frostbite were severely punished. Even in the harsh winter of 1941–42 the Russians were able to spend many days in the snow, protected only by simple windbreaks, without detriment to their health.

For instance, in the winter of 1941–42, from 6 December 1941 to 14 April 1942, the Soviets carried out their first major offensive in the area west of Moscow. They penetrated to the Vyazma-Rzhev line. In their large-scale attack from the Stalingrad-Don area the Russians succeeded in encircling the German Sixth Army at Stalingrad, and in overrunning the front of the German allies. Between 19 November 1942 and the middle of March 1943 they succeeded in creating a fluid situation along a 1,000-mile stretch of the Eastern Front, and in penetrating up to 300 miles toward the west.

On Christmas of 1943 the Russians began an offensive from the area southwest of Kiev. It continued until the thaw started in March 1944 and led to the annihilation of German divisions in the Cherkassy pocket, the encirclement of Tarnopol and of the German First Panzer Army, and the retrograde correction of the German front to a line east of Stanislaw-Lwow-Kowel.

The Russian general offensive, begun along the entire Eastern Front in mid-January of 1945, culminated in the occupation of Hungary, Poland, Silesia, East Prussia, and Pomerania. After short local halts, the Russians stood before Vienna and Berlin, and in cooperation with the armed forces of the Western Powers brought about the end of the war.

II. The Employment of Mass

Only the use of masses of men and equipment enabled the Russians to accomplish these great feats. Even in World War I the "steam roller" was the core of the Russian system of attack. At that time the concept was limited to the employment of the human mass, which resulted in initial successes but cost myriads of human lives. Later, undermined politically from within, the system finally disintegrated sealing the doom of the Russian Imperial Army.

The surprise element of the attack on the Eastern Front by the Germans in June 1941 prevented the immediate use of Russian mass tactics and caused Soviet reverses. Relying, however, on inexhaustible reserves and the rigors of its boundless territory, the Soviet Army concentrated human masses and created anew the famous steam roller. Strengthened by a mass of equipment and welded together by the caresses and whiplashes of communism, it became invincible despite numerous setbacks. Step by step the German Army was pounded to pieces and crushed as it succumbed to what might be termed the modern "super steam roller."

As early as 1941 the Russians made repeated attempts to check the German offensive by counterattacks. To this end they opposed it with cavalry and tank forces, and sometimes with masses of infantry. They succeeded in checking the German Army locally and temporarily, but were unsuccessful in stopping the great offensive as a whole. The following examples illustrate the Russian type of attack at that time.

The battle in the Dnepr-Berezina triangle (mid-July to mid-August 1941) resulted from Timoshenko's plan to envelop in a pincers movement both flanks of the German armored units which had advanced across the Dnepr, and to cut them off from the rear. Since the northern arm of the pincers had already been destroyed near Orsha, and the German LIII Infantry Corps had almost made connection with the armored units on the Dnepr by a rapid advance via Slutsk-Bobruysk, a battle developed not between Timoshenko and the armored units, but between the Russians and LIII Infantry Corps, which was assuming the protection of the flanks and rear of the German armored unit. (Map 3)

The attack was very carefully prepared. As early as February 1941, during a conference near Bobruysk, Timoshenko had discussed this operation in a kind of map exercise with his higher commanders. In order to put it into effect, three divisions located in the vicinity of Kiev were transferred to the neighborhood of Gomel immediately after the outbreak of hostilities. Here, they were excellently concealed and kept in readiness until the time for action came. The Moscow School of Artillery was called upon. By interrogation of Russian prisoners, 53 different rifle regiments were identified during the course of battle, some of them coming from as far away as the Caucasus. In order to take immediate advantage of the expected success, a detachment of specialists from Moscow very quickly repaired the railroad bridge over the Dnepr at Zhlobin during the course of the battle. The bridge had been blown prematurely when German armored units approached the town. Timoshenko reopened the bridge in person. Thus, all preparations seemed to have been made to ensure the success of the operation and to exploit it. The higher command had done as much as could be expected.

Tactical measures during the first stage of the battle were entirely appropriate. The German 4th Panzer Division, which had crossed the Dnepr near Rogachev, was thrown back across the river. Simultaneously with the withdrawing armored vehicles, the Russians crossed the Dnepr in close pursuit, and for the time being there was nothing to stop their advance, since the armor had turned off to the north. At Zhlobin, where there were only weak armored security forces, the Soviet Army faced practically no opposition at all. The Russians did not take advantage of this favorable situation. They moved forward very slowly, so that very small, hastily motorized forces of the German 255th and 267th Infantry Divisions could be thrown against them. These German forces held the Russians at bay until the divisions themselves arrived after a day or two. Thus it happened that the Russians pushed only 6 miles to the west of the Dnepr both at Zhlobin and Rogachev. One might almost think that this was the area that had been planned as the first day's objective in Timoshenko's war game. More territory could unquestionably have been taken, had the advance been conducted more energetically. Lack of follow-up, however, can be attributed to the intermediate command. By the time the high command learned of the situation, it was too late. The subsequent Russian attacks were carried out in the form of outflanking and enveloping movements while pinning down the front. They occasioned many a critical situation, but when the German's greatest worry had been eliminated after the commitment of the 55th Infantry Division, and the front began to be more stabilized, the Russian command exhausted itself by repeated attacks on the same points. The Russians failed to recognize that the opposing forces were now of equal strength, and that therefore nothing more was to be achieved by the battle.

It was soon evident that the strategic situation of the Russians was becoming steadily less tenable. The advance of XII Infantry Corps northeast of Rogachev, and of XLIII Infantry Corps to the lower Berezina, made the grave danger facing the Russians unmistakably clear. Nevertheless, they held firmly to the originally planned thrust, even after it was time to escape envelopment. This situation led to the formation of the Gomel pocket and the annihilation of the Russian main attack army, a catastrophe which could have been avoided, had the Russian command drawn its conclusions from the situation in time.

In the fighting in the Porechye bridgehead on the Luga (mid-July to August 1941) a strong combat element of the 6th Panzer Division succeeded in effecting a surprise capture of the two Luga bridges at Porechye (60 miles southwest of Leningrad) on 14 July 1941, and in forming a bridgehead. (Map 4) The strong armored forces which were to follow remained stuck for days in swampy forests. Therefore, the 6th was for a long time entirely on its own. Three alerted proletarian divisions and armored units were dispatched from Leningrad to the Luga by rail and motor vehicles, with the mission of destroying the German force consisting of two infantry battalions, one panzer battalion, two artillery battalions, one Flak battalion, and a company of engineers.

Nevertheless, the German force was able to stand its ground against an enemy six times as strong, despite the fact that it could no longer be reached and supported by aircraft. The bridgehead was over 2 miles deep, but only 400 yards wide, and was completely surrounded by swampy forests which could be traversed in summer by infantry. The Russians, however, had no visibility, because the edges of the woods were in German hands. Russian artillery fired 20 times as many rounds per day as did the German batteries. It pounded the bridges and the edges of the woods. In unobserved fire the Russians sent more than 2,000 medium shells a day in the direction of the bridges without ever hitting them. Enemy aircraft roared all day above the narrow corridor of the bridgehead and inflicted serious losses. Soon, therefore, the road was lined with a long row of German soldiers' graves, marked with birchwood crosses. As often as 10 times a day the enemy attacked the road fork which was enclosed by the projecting arc of the bridgehead. Each attack was headed by as many tanks, echeloned in depth, as the narrow road would accommodate. Time and again the enemy attacks were repulsed, and time and again they were renewed. Wave after wave of Russian forces assembled, concealed by many wrecked tanks and heaps of corpses, and stormed recklessly into the murderous defensive fire. The attacks did not subside until the enemy no longer had the necessary men and ammunition at his disposal. Soon, however, replacements of proletarians and new ammunition supplies arrived from Leningrad. The assaults on the road fork were stubbornly and incessantly resumed until passage through the narrow attack corridors was no longer possible because they were completely clogged by disabled enemy tanks and decaying heaps of corpses. Later, the Germans counted more than 2,000 dead Russians and 78 knocked-out enemy tanks in this narrow combat zone alone.

The only variation consisted of enemy attacks, sometimes along both sides of the road leading to the fork from the north, and at other times on the road coming from the south. A simultaneous attack from both sides never took place, however, for different divisions were involved which did not coordinate their attacks. The tanks always attacked along the northern road. Their attacks, too, ceased only when it was no longer possible to pass between the 78 wrecks. The swampy forests and a reservoir 50 feet deep prevented bypassing.

The next attacks were made through the woods to cut off the bridgehead. It was impossible to prevent the mass assaults since there were only a small number of forces to man the long flanks of the bridgehead. The attacks were expected and the defense arranged accordingly. Two tank units held in readiness, one at the road fork and one at the bridges, were reinforced by armored infantry companies and had orders to attack immediately from both sides and to annihilate the enemy if he broke through to the road. This flexible fighting method proved effective. With the same stubbornness he had formerly shown in attacking the road fork, the enemy made several assaults daily on the flanks, and each time succeeded in reaching the road. Within a scant half hour he would be attacked from both sides and destroyed by tank fire. Only remnants succeeded in escaping through the woods. Not until the attempt to cut off the group failed repeatedly with severe losses, did the enemy give up this procedure and attack the bridges directly. In this way a 150-man Russian battalion succeeded in reaching the southern bridge after overcoming its weak defenses. While the forces were crossing the bridge, German armored personnel carriers took them by surprise and completely destroyed them. Contrary to his usual custom, the enemy did not repeat this venture. Instead, he sent a reinforced infantry regiment from Kingisepp, and later two battalions of the 3d Proletarian Division, again the rear of the bridgehead. Both operations were caught in a pincers movement between the leading elements of the main German forces, which had been stalled but were then approaching from the southwest, and a reinforced armored unit of the bridgehead coming from the opposite direction. The regiment was scattered and the two battalions were destroyed. The bridgehead was then strengthened by German forces coming up from the rear. and became impregnable.

The foregoing example demonstrates the complete failure of the intermediate and lower Russian commands which did not understand how to coordinate the various units and weapons into a common, simultaneous attack from all sides. The weak German units, which were fighting a losing battle, would not have been able to withstand such an attack.

The Russian attacks in the battle southeast of Plavskoye (Map 5) in November 1941 developed from a pure meeting engagement. The

Russian forces, which were to advance against German armored forces before Tula, suddenly faced LIII Infantry Corps coming from the west. The Russian command adapted itself skillfully to the new situation—a situation made easier on both sides because the points met first and each opponent was then able to deploy its forces from depths. It was, therefore, several days before the battle mounted to a climax once both sides had deployed their long columns.

Strategically, the Russian thrust on Tula was well planned. At this stage the Russian higher command also acted correctly from a tactical standpoint, and the Russian 6th Rifle Division, which bore the initial brunt of the battle, was tactically well commanded. Under the completely unclarified conditions, the Russian higher command apparently wanted to avoid having its foremost division repulsed, and so brought it to a halt. The division, deployed laterally, blocked the advance of the first German regiment to appear. In addition, the Russian higher command attempted to bring up for the anticipated battle all forces available in the vicinity. The Russian cavalry division, which had advanced north of the combat area, abandoned its operations against the 112th Infantry Division and appeared on the scene. A cavalry division located near Yefremov was ordered forward, over side roads past the march columns of the Russian rifle division, to cut the Orel-Plavskoye road in the neighborhood of Chern. In this connection there ensued a sharp clash between the commander of the Bryansk Army Group in Yelets, who was in command here, and the commander of the cavalry division from Yefremov.

According to captured documents the army group commander ordered the cavalry division to proceed by a forced march to the Orel-Tula road, a distance of 40 miles. The cavalry division commander made a written reply stating that this was impossible because 20 percent of his horses were completely unshod and, of the remaining horses, 80 percent had no hind shoes and 20 percent no fore shoes. For that reason the division was unfit to move. Thereupon the higher commander repeated the order for the forced march. The cavalry division started out, but actually went only 25 miles. It could go no farther, but it had effected a connection with the front and extended it. The iron determination of the army group commander had not achieved the desired success, but it had created new difficulties for the Germans.

The Red 6th Rifle Division was very mobile in its fighting. When its western flank was threatened and the divisions following it had not yet arrived, it withdrew far enough in its numerous trucks to eliminate the threat to its flank. It did not move forward again until the next division could be turned to cover the threat to its flank. The other divisions were employed according to the prevailing situation and formed concentrated points of main effort. They attacked by deploying directly from march columns, without going into assembly positions. Not until they were all disposed along the battle front was the tank brigade, which had been far to the rear, committed in order to strike the decisive blow against whatever seemed to be the weakest spot. This was on about the fifth day of battle. Only rapid shifting of German forces prevented a Russian success.

To be sure, when LIII Infantry Corps subsequently launched its attack, the rifle division, which had become immobile to a great extent, was wiped out and the westernmost cavalry division had already been eliminated. However, the bulk of the Russian forces escaped annihilation by hasty withdrawal, sacrificing much of their equipment. This was the first instance in the corps' experience in which the Russian high command withdrew troops from a situation that was beginning to take an unfavorable turn, instead of continuing to attack. The intervention of LIII Infantry Corps upset the Russian large-scale plan of operation from the very start. The attempt to force a way to Tula in spite of this had failed. The Russian high command now actually drew the proper inferences from the situation. They adopted new methods.

III. Development of Russian Offensive Tactics

The Russians carried out their first preconceived large-scale offensive in the severe winter of 1941–42 in the Moscow area. It was well thought out, and cleverly exploited the detrimental effect which the muddy period and the onset of winter had on the striking power of the German Army, unprepared as it was for winter warfare. This offensive marked the turning point in the Eastern Campaign; it did not, however, decide it, as the Russians had expected it to do. Therefore, it did not achieve the intended purpose.

According to the Russian fighting method the mass attack was supposed to shatter the German front. Units penetrating and infiltrating through the lines were to cut off the supply lines.

The German front was not yet ready for defense when frost set in. Some sectors were still fluid. The solidly frozen ground and the exceedingly heavy snowfalls precluded the methodical construction of a defensive position. Taking advantage of the dusky weather and the blizzards, the first waves of Russians, clad in white camouflage coats, worked their way close to the German positions without being detected. Wave after wave, driven on by the commissars, surged against the German lines. At that time the Russians knew nothing about methodical preparation for an attack by concentrated fire of heavy weapons and artillery, or by the employment of massed tanks.

When the attacks continually failed and enormous losses were incurred, the Russians changed to infiltration tactics. Forces capable of rapid movement were generally used for this purpose. Tanks and ski units were preferred when the terrain permitted their use. Except for a few local reconnaissance thrusts, the Russians regularly attacked on a broad front. They always assembled far superior forces for those attacks. Cooperation between the different arms of the service improved noticeably. It was patterned after German offensive tactics. Russian methods of attack were subjected to many changes as a result of war experiences. In 1941–42 the Russians always resorted to mass attacks after bringing up strong reserves. Thus, for example, they repeated their attacks in the same place against the Fourth Army for weeks at a time during the so-called Battles of the *Rollbahn* (express motor highway, in this case the Minsk-Moscow road) near Smolensk between 21 October and 4 December. Cooperation between the various infantry weapons likewise was imperfect. Attacks on German unit boundaries, which the Russians always sought and usually detected, were dangerous. Artillery support was active, but as yet often without a definite fire plan.

The year 1943 brought a definite change in the method of attack. Concentrated artillery fires were employed more frequently and supplemented by massed mortar attacks. The Russians tried to infiltrate through known German weak points. For this purpose they preferred forest areas or hollows previously designated by the tactical command. If they succeeded in infiltrating by this system, they immediately entrenched themselves and laid mines. Subsequently, a period of vulnerability set in because the artillery and heavy weapons were brought up slowly, and cooperation with them ceased abruptly.

The employment of massed tanks brought about a revolutionary change in Russian tactics in 1944. After a drum fire of artillery, a large number of tanks led off the attack, followed by the infantry in deep wedges. While the artillery gave good support at first, communications with it frequently broke off during a further advance. To the very end of the war it was difficult for the Russians to coordinate fire and movement. The penetrations were deep, and invariably in a straight line. Then a halt was called in order to bring up the greatest possible number of infantry during the night. These masses of infantry dug in as soon as they reached the points of the attack. The assault wedges closed up in echelons behind the tanks.

Since the German counterattacks were usually launched when the enemy infantry was separated from its tanks, the Russians began in 1945 to make deep thrusts with infantry riding on tanks. These thrusts often went so deep that contact with the main forces was lost. The Russians were able to take the risk because the German front of 1945 no longer had adequate reserves available to destroy the far-advanced, strong enemy forces.

Though the Russians built field fortifications whenever they halted it did not follow that they had plans of attacking. The Russian always dug in. The time to be suspicious came only when a gradual sapping toward the German lines into a jump-off position for an assault could be noted in connection with their entrenchment activities. That usually meant preparation for attack. Recognized preparations behind a front sector did not necessarily indicate an attack at that particular point. Surprise attacks were launched by skillfully and quickly shifting attack forces to the planned assembly area during the course of one night. Numerous reconnaissance thrusts, supported by artillery and tanks, and conducted on a broad front both by day and night in strength up to a regiment, were to procure information for an attack and to confuse or deceive the Germans as to the time and place of the attack. Movement behind the front, even at night, was not necessarily followed by an attack. The Soviets were very skillful in the use of feints, sham installations, and dummy matériel of all types. Evaluation of artillery observation data, often painstakingly carried on for weeks, and constant interrogation of prisoners, whose statements were checked by German reconnaissance operations, produced reliable evidence of an impending attack. The Russians often cleverly concealed a projected assembly of their artillery by extraordinary emplacement activity and by a highly mobile employment of roving guns and batteries. They were also very cautious about fire for registration whenever it was not executed for purposes of deception.

Not until later did the Russians make extensive use, in the attack, of artillery fire based on mathematical computation. Despite all their efforts to conceal their true intentions, however, the pattern of enemy artillery activity, carefully worked out day by day, still revealed very reliable clues as to impending enemy attacks. The Germans could often observe that a few days before an attack the enemy moved about as little as possible by day or night, and that his combat activity decreased noticeably, until suddenly the attack was launched out of a clear sky.

Secondary attacks and feints were often launched at the same time as the main attack in order to make the assult front appear as broad as possible, and at the same time cause the Germans to split up their defense forces. In the summer of 1943, the Russians used smoke on a broad front while carrying out attacks across the Donets. This concealed preparations and denied observation to the Germans. At that time the Russians incurred severe losses. The German XXX Infantry Corps repulsed all attacks and attempts to cross the Donets by immediately concentrating the fire of all heavy weapons straight into the smoke. At that time the Russians still were very inexperienced in the use of smoke and did not use dummy smoke screens.

Likewise, they failed to understand the principle of laying a smoke screen over German observers in order to blind them. At that time, too, the Russians did not necessarily carry out their attacks with artillery preparation. When they did, their artillery, massed into points of main effort, laid a rigid concentration on infantry positions, battery positions, towns, and road junctions. During the artillery preparation the infantry worked its way forward into the jump-off position by infiltration, and from that point made a mass advance. When the first objective, which was still within range of the supporting artillery, was reached, a long halt was called since the Russians were not in a position to displace their artillery and heavy weapons forward in a manner that would allow a continuous forward thrust. The infantry immediately dug in and felt its way forward only by combat reconnaissance. During the halts, the infantry had to rely almost exclusively on local support from accompanying tanks and mortars. The Russian heavy mortar battalions (probably 120-mm.) were an ideal direct-support artillery for infantry. However, they also were too slow for continuous support of the infantry in an attack. The infantry-support tanks acted very cautiously and fought more in the manner of self-propelled assault guns, or like armored artillery pieces of the infantry.

Even at the beginning of the Eastern Campaign the Russian infantry was very clever at utilizing terrain features. If the Russians could not continue their current main attacks with the desired success by day, they proceeded to launch local attacks at night. In that case, they either launched sudden mass attacks, or infiltrated at many points through the German lines which for the most part were lightly manned. Thus, in a night attack on the 97th Light Infantry Division in the winter of 1941, they broke through east of Artemovsk in the Donets Basin with an entire cavalry division. However, the next day this division was cut down to the last man. Also in the battle of encirclement at Uman in 1941, and at Beli—southeast of Toropets—in November 1942, thousands of Russian soldiers without equipment or heavy weapons penetrated the German lines during hours of darkness by piecemeal infiltration. Well-prepared night attacks were rare.

Low-flying aircraft, supporting the main efforts, picked as their principal targets the defending infantry, batteries, reserves, supplies, and villages in the divisional combat sectors. The attack planes did not venture far into the rear area.

In order to develop the war of position into mobile warfare later in the Eastern Campaign, the Red Army command concentrated its forces, which were numerically far superior to those of the Germans, into more and more powerful masses at the points of main offort and, after heavy artillery concentrations, broke through or sent the German front reeling. The Russian command attempted to conceal strategic preparations for large-scale attacks from German reconnaissance, and therefore carried them out only at night. Preparations could thus be detected only by night aerial reconnaissance. Night truck transport operations on a grand scale (2,000 to 3,000 trucks in each direction in the course of one night), which usually took place shortly before a large-scale attack, were the first reliable indication of an imminent Russian offensive.

IV. The Use of Armor

The Russian armored force played only a subordinate role at the beginning of the war. In the advance of 1941, most German troops encountered only small units which supported the infantry in the same manner as the German self-propelled assault guns. The Russian tanks operated in a very clumsy manner and were quickly eliminated by German antitank weapons. The Russians carried out counterattacks with large tank forces, either alone or in combined operations with other arms, only at individual, important sectors.

On 23 June 1941 the German Fourth Panzer Group, after a thrust from East Prussia, had reached the Dubysa and had formed several bridgeheads. The defeated enemy infantry units scattered into the extensive forests and high grain fields, where they constituted a threat to the German supply lines. As early as 25 June the Russians launched a surprise counterattack on the southern bridgehead in the direction of Raseiniai with their hastily brought-up XIV Tank Corps. They overpowered the 6th Motorcycle Battalion which was committed in the bridgehead, took the bridge, and pushed on in the direction of the city. The German 114th Armored Infantry Regiment, reinforced by 2 artillery battalions and 100 tanks, was immediately put into action and stopped the main body of enemy forces. Then there suddenly appeared for the first time a battalion of heavy enemy tanks of previously unknown type. The tanks overran the armored infantry regiment and broke through into the artillery position. The projectiles of all defense weapons (except the 88-mm. Flak) bounced off the thick enemy armor. The 100 tanks were unable to check the 20 enemy dreadnaughts, and suffered losses. Several Czech-built tanks (T36's) which had bogged down in the grain fields because of mechanical trouble were flattened by the enemy monsters. The same fate befell a 150-mm. medium howitzer battery which kept on firing until the last minute. Despite the fact that it scored numerous direct hits from as close a range as 200 yards, its heavy shells were unable to put even a single tank out of action. The situation became critical. Only the 88-mm. Flak finally knocked out a few of the Russian KV 1's and forced the others to withdraw into the woods.

One of the KV1's even managed to reach the only supply route of the German task force located in the northern bridgehead, and blocked it for several days. The first unsuspecting trucks to arrive with supplies were immediately shot afire by the tank. There were practically no means of eliminating the monster. It was impossible to bypass it because of the swampy surrounding terrain. Neither supplies nor ammunition could be brought up. The severely wounded could not be removed to the hospital for the necessary operations, so they died. The attempt to put the tank out of action with the 50-mm. antitank gun battery, which had just been introduced at that time, at a range of 500 yards ended with heavy losses to crews and equipment of the battery. The tank remained undamaged in spite of the fact that, as was later determined, it got 14 direct hits. These merely produced blue spots on its armor. When a camouflaged 88 was brought up, the tank calmly permitted it to be put into position at a distance of 700 yards, and then smashed it and its crew before it was even ready to fire. The attempt of engineers to blow it up at night likewise proved abortive. To be sure, the engineers managed to get to the tank after midnight, and laid the prescribed demolition charge under the caterpillar tracks. The charge went off according to plan, but was insufficient for the oversized tracks. Pieces were broken off the tracks, but the tank remained mobile and continued to molest the rear of the front and to block all supplies. At first it received supplies at night from scattered Russian groups and civilians, but the Germans later prevented this procedure by blocking off the surrounding area. However, even this isolation did not induce it to give up its favorable position. It finally became the victim of a German ruse. Fifty tanks were ordered to feign an attack from three sides and to fire on it so as to draw all of its attention in those directions. Under the protection of this feint it was possible to set up and camouflage another 88-mm. Flak to the rear of the tank, so that this time it actually was able to fire. Of the 12 direct hits scored by this medium gun, 3 pierced the tank and destroyed it.

The Russian had not taken advantage of the critical situation of the German division which had resulted from the employment of the heavy 65-ton tanks. His infantry, which had broken through, did not become active again, but passively watched the proceedings. Therefore, it was possible to withdraw strong forces from the northern bridgehead and send them against the rear of the attacking tank corps. The latter immediately abandoned its success and retired to the east bank of the river. There, despite the fact that strong neighboring armored forces had already enveloped it and were attacking from the rear, it again held out far too long. The result was that the Russian tank corps lost the bulk of its tanks in the swamps. The infantry scattered and made its way through the woods along swamp paths.

On 26 June 1941 the Russians, by a tank corps thrust, wanted to relieve their forces which were encircled near Rawa Ruska, north of Lwow. This tank corps consisted only of tank chassis mounting machine guns and guns up to 150-mm.; it had no motorized infantry support. Near Magierow it encountered organized defense of the German 97th Light Infantry Division in a day and night attack, and was repulsed. Sixty-three Russian armored vehicles were knocked out.

In the thrust on Vyazma (early in October 1941) the German 6th Panzer Division, committed in the main effort, reached the upper Dnepr and captured the two bridges there by a *coup de main*. (Map 6) That maneuver cut off Russian forces which were still west of the river and assured a continued thrust to the east.

On the following day the Russians attempted to parry this severe blow by a flank attack. One hundred tanks drove from the south against the road hub of Kholm. They were, for the most part, medium tanks, against which the Germans could send only 40 light tanks and 1 armored infantry company. However, these weak forces were sufficient to contain the dangerous thrust until antiaircraft and antitank guns could be organized into an adequate antitank defense between Kholm and the southern Dnepr bridge. Split up by tanks in forest fighting, the Russians never succeeded in making a powerful, unified tank thrust. Their leading elements were eliminated as they encountered the antitank front. As a result, the Reds became even more timid and scattered in breadth and depth in such a way that all subsequent tank thrusts, carried out in detail by small groups, could be met by the German antitank front and smashed. Kholm and the Dnepr bridge, as well as their connecting road—which the Russians had already taken under intermittent tank fire—remained in German hands. After 80 Russian tanks had been put out of action, a break was made through the strongly fortified position on the east bank of the Dnepr, which was occupied by Red reserves, and the thrust from the southern bridgehead continued without concern for the Russians. The flank attack in detail by 100 Russian tanks near Kholm succeeded in delaying, but not in stopping, the advance of the German 6th Panzer Division.

While the German division with all its combat elements rolled along a road deep into Russian territory, hastily assembled Russian tank units and infantry, supported by several batteries, attempted to attack the 25-mile flank of the march column and stop its advance. Some of the Soviet batteries remained in their former positions and merely turned their guns around, while others rushed up at full speed and assumed fire positions in the open. Infantry and tanks advanced in a widespread chain against the German column, and the artillery immediately opened fire with every battery as soon as it had shifted its front.

The attack turned out badly for the Russians. In an instant the German division was firing all its weapons. The division resembled a mighty battleship, smashing all targets within reach with the heavy caliber of its broadsides. Artillery and mortar shells from 300 throats of fire hailed down on the enemy batteries and tanks. Soon the Soviet tanks were in flames, the batteries transformed into smoking heaps of rubble, and the lines of skirmishers swept away by a swath of fire from hundreds of machine guns. In 20 minutes the work of destruction was completed. The advance continued and on the same day reached Vyazma, its objective. This completed the encirclement of 400,000 Russians.

The Russian clearly recognized and twice tried to prevent the division's intent to break out of the bridgehead and push eastward from Kholm. He failed each time, although he had adequate forces and means at his disposal. In both cases the Russian command was at fault. In the first instance it was unable to execute a coordinated blow with a force of 100 tanks; in the second, it did not succeed in readying and coordinating all the available forces and antitank defenses in time.

The command of large tank units was usually difficult for the Russians even in later years. They had only a few competent armored commanders. In the tank force, too, successes were achieved only by the reckless use of masses. However, those tactics failed whenever even relatively adequate defense means were available to the Germans.

In the winter of 1942–43 the Russians employed four tank corps for a break-through near Kamenka on the Donets. By a thrust on Voroshilovgrad, the Russians would have been able to strike a crushing strategic blow at the deep left flank of Army Group von Manstein on the south bank of the Donets. Instead, they were attracted by Kamenka and Millerovo and therefore made an assault only against the wings of a provisional army which were strongly defended at that point, and in spite of their superiority achieved only a tactical success. At that time the attack of the Russian main force came as a surprise. In minor thrusts launched during the preceding days the Russians had probed the front of the provisional army, which consisted only of separate strong points (a frontage of 120 miles, and held by one infantry division, one reinforced mountain infantry regiment, one SS regiment reinforced by armor, one panzer battalion, and several Flak batteries). The Russians proved that they were still unable at that time to employ large tank units strategically.

During the course of the well-commanded Russian counteroffensive of Byelgorod, massed Russian tanks reached the area around Bogodukhov, northwest of Kharkov, and Graivoron on the first day (5 August 1943), and then flowed like lava into the broad plain east of the Vorskla, where they were halted by German counteroperations from the Poltava-Akhtirka area. (Map 7)

Kharkov constituted a deep German salient to the east, which prevented the enemy from making use of this important traffic and supply center. All previous Russian attempts to take it had failed. Neither tank assaults nor infantry mass attacks had succeeded in bringing about the fall of this large city. Boastful reports made by the Russian radio, and erroneous ones by German pilots, announcing the entry of Russian troops into Kharkov at a time when the German front stood unwavering, did not alter the facts. When the Russian command perceived its mistake, Marshal Stalin ordered the immediate capture of Kharkov.

The rehabilitated Russian Fifth Tank Army was assigned this mission. The German XI Infantry Corps, however, whose five divisions firmly sealed off the city in a long arc, recognized the new danger in time. It was clear that the Russian Fifth Tank Army would not make a frontal assault on the projecting Kharkov bastion, but would attempt to break through the narrowest part of the arc west of the city, the so-called bottleneck, in order to encircle Kharkov. Antitank defenses were installed at once. All available antitank guns were set up on the northern edge of the bottleneck, which rose like a bastion, and numerous 88-mm. Flak guns were set up in depth on the high ground. The antitank defense would not have been sufficient to repulse the expected mass attack of Russian tanks, but at the last moment the requested 2d SS Panzer Division ("Das Reich") arrived with strong armored forces and was immediately dispatched to the sector most endangered.

The 96 Panthers, 32 Tigers, and 25 self-propelled assault guns had hardly taken their assigned positions when the first large-scale attack of the Russian Fifth Tank Army got under way. The first hard German blow, however, hit the assembled mass of Russian tanks which had been recognized while they were still assembling in the villages and the flood plains of a brook valley. Escorted by German fighters, which cleared the sky of Russian aircraft within a few minutes, wings of heavily laden Stukas came on in wedge formation and unloaded their cargoes of destruction in well-aimed dives on the assembled tanks. Dark fountains of earth erupted skyward and were followed by heavy thunderclaps and shocks which resembled an earthquake. These were the heaviest, 2-ton bombs, designed for use against Russian battleships, which were all the Luftwaffe had to counter the Russian attack. Wing after wing approached with majestic calm, and carried out its work of destruction without interference. Soon all the villages occupied by Soviet tanks were in flames. A sea of dust and smoke clouds illuminated by the setting sun hung over the brook valley. Dark mushrooms of smoke from burning tanks, victims of the heavy air attacks, stood out in sharp contrast. The gruesome picture bore witness to an undertaking that left death and destruction in its wake. It had hit the Russian so hard that he could no longer launch the projected attack on that day, in spite of Stalin's order. A severe blow had been inflicted on the Russians, and the time needed for organizing German measures had been gained.

The next day the Russians avoided mass grouping of tanks, crossed the brook valley at several places, and disappeared into the broad cornfields which were located ahead of the front, but which ended at the east-west main highway several hundred yards in front of the main line of resistance. During the night motorized infantry had already infiltrated through the defense lines in several places and made a surprise penetration near Lyubotin into the artillery position. After stubborn fighting with the gun crews, 12 howitzers without breechlocks—which the crews took with them—fell into Russian hands. The points of the infiltrated motorized infantry already were shooting it out with the German local security in the wood adjoining the corps command post.

During the morning Red tanks had worked their way forward in the hollows up to the southern edges of the cornfields. Then they made made a mass dash across the road in full sight. The leading waves of Russian T34's were caught in the fierce defensive fire of the Panthers, and were on fire before they could reach the main line of resistance. But wave after wave followed, until they flowed across in the protecting hollows and pushed forward into the battle position. Here they were trapped in the net of antitank and antiaircraft guns, Hornets (88-mm. tank destroyers), and Wasps (self-propelled 105-mm. light field howitzers), were split up, and large numbers of them put out of action. The last waves were still attempting to force a break-through in concentrated masses when they were attacked by Tigers and self-propelled assault guns, until then mobile reserves behind the front, and were repulsed with heavy losses. The first thrust of the Russians was repelled. The price they paid for this mass tank assault amounted to 184 knocked-out T34's.

In the meantime, German infantry reserves supported by self-propelled assault guns from the 3d Panzer Division had captured the lost battery positions together with all pieces and, west of Lyubotin behind the main line of resistance, had bottled up the battalion of infiltrated enemy motorized infantry. Stubbornly defending themselves, the Russians awaited the help that their radio had promised.

The Russian changed his tactics and the next day attacked farther east in a single deep wedge, using several hundred tanks simultaneously. But even while they moved across open terrain along the railroad, numerous tanks were set on fire at a range of 2,000 yards by the long-range weapons of the Tigers and Hornets. The large-scale Red attack was not launched until late in the forenoon. As the tanks emerged from the cornfields this time, they were assailed by the concentrated defense of all Tigers, Hornets, Panthers, self-propelled assault guns, and antiaircraft and antitank guns, and the attack collapsed in a short time with the loss of 154 tanks. The weak rifle units which followed were mowed down by the concentrated fire of German infantry and artillery as they emerged from the cornfields. The encircled Red motorized battalion had waited in vain for aid, but continued to fight on with incredible tenacity. In the late afternoon its radio announced the defeat of the unit and then fell silent forever. After 48 hours of heroic defense, the Red battalion was killed to the last man, including radio operators.

The losses thus far incurred by the Russians were enormous. However, they still possessed more than a hundred tanks, and experience had taught the Germans that further attacks were to be expected, even though they were predestined to failure in view of the now vastly superior defense. The few tankers taken prisoners were aware that death, or, if they were lucky, capture, awaited every one of their comrades.

Contrary to all expectations, an eerie calm prevailed throughout the following day. Several Red tanks crawled about in the cornfields and towed the damaged tanks away in order to reinforce their greatly depleted ranks. Summer heat shimmered over the bloody fields of the past days of battle. A last glow of sunset brought the peaceful day to a close. Might the enemy have given up his plan, or even refused to obey the supreme order to repeat the attack?

He came back, and on the same day. Before midnight, considerable noise from tanks in the cornfields betrayed his approach. The enemy intended to achieve during the night what he had failed to gain by daylight attacks.

Before he had reached the foot of the elevated terrain, numerous flashes from firing tanks had ripped the pitch-black darkness of the night and illuminated a mass attack of the entire Russian Tank Army on a broad front. Tanks knocked out at close range already were burning like torches and lighting up portions of the battlefield. More tanks joined them. The German antitank guns could no longer fire properly, since they could hardly distinguish between friend and foe; German tanks had entered the fray, ramming Russian tanks in a counterthrust or piercing them with shells at gun-barrel range in order to block the break-through. A steady increase in the flash and thunder of tank, antitank, and antiaircraft guns could be perceived after midnight. The main force of the German tanks had launched a counterattack. Many tanks and several farm buildings went up in flames.

The plateau on which this great night tank duel was fought was illuminated by their pale light. This made it possible to recognize the contours of Red tanks at a distance of more than 100 yards, and to shell them. The thunderous roll turned into a din like the crescendo of kettledrums as the two main tank forces clashed. Gun flashes from all around ripped the darkness of night throughout an extensive area. For miles, armor-piercing projectiles whizzed into the night in all directions. Gradually the pandemonium of the tank battle shifted to the north. However, flashes also appeared farther and farther behind the German front, and fiery torches stood out against the night sky. Not until 2 or 3 hours later was calm restored in the depth of the German front. The conflict also gradually subsided in the battle position.

After daybreak the Germans could feel the battle was won although there were still Red tanks and motorized infantry in and behind the German position, and here and there a small gap still remained to be closed. The mopping up of the battle position, however, lasted all morning. By noon the position was in German hands and again ready for defense. Only a small patch of woodland, close behind the main line of resistance, was still occupied by Red motorized infantry supported by a few tanks and antitank guns. All attempts to retake this patch of woods had failed with heavy German losses. Even heavy, concerted fires of strong artillery units could not force the Russians to yield.

The tenacious resistance was ended only by an attack of flame-throwing tanks, which burned the entire strip of woods to the ground. The foremost of the Red tanks which had made the deep forward thrust was captured at the western outskirts of Kharkov by a divisional headquarters, and the crew members were taken prisoner. All the rest were put out of action by Flak teams.

The Red plan to take Kharkov by a large-scale night attack of the entire tank army had failed. The losses were more than eighty burned-out tanks, many hundreds of dead, thousands of wounded, and a considerable amount of equipment in this night of battle. The Russian Fifth Tank Army in the effort to recapture Kharkov lost 420 tanks in 3 days of fighting, and suffered such heavy losses of men and equipment that it ceased to be a combat factor for the foreseeable future. Kharkov remained in German hands until the high command ordered the troops stationed there to retire.

Blunders on the part of the leaders were only partially responsible for the fact that every one of the Red tank attacks failed, although the troops fought with extraordinary bravery. It was striking that the enemy had only weak infantry and artillery forces, and that his air forces did not participate effectively enough in operations. For these reasons the tank forces could not be adequately supported and their successes could not be exploited. The Fifth Tank Army seems to have been forced to premature action for reasons of prestige by orders of the Russian Supreme Command.

In the winter of 1943–44 the German XXX Infantry Corps' 16th Panzer Grenadier Division experienced a break-through of strong Russian tank forces with a long-range objective in the Dnepr bend south of Dnepropetrovsk. Here the Russians, with tank divisions followed by motorized forces, made a deep thrust against the left flank of the German Sixth Army forces which were withdrawing in front of Nikopol. At that time, the German front enclosed Nikopol in a semicircle east of the city. As this strong tank thrust gained in depth, it decreased in power because it split up. It did not achieve its strategic objective. In spite of the fact the Russian forces were many times superior in number, the German Sixth Army succeeded in containing the thrust in the depth of its sector and in forming new fronts. Even after a double envelopment by far superior tank forces (nine tank corps), the Russians, after encircling the German Sixth Army in Bessarabia, did not succeed in blocking the road over the Transylvanian Alps against the remnants of the Sixth and Eighth Armies. In the tank battle of Debrecen (summer 1944), too, Russian tanks and motorized units split up in such a manner, without being reconcentrated, that weak German panzer divisions succeeded not only in preventing a break-through but also in throwing the Russians back again toward Debrecen. The leadership of these large, strategic armored forces was inadequate. In this instance only the enemy's enormous numerical superiority, and his mobility, brought him local successes.

Even in the last months of the war the Russians committed blunders in the command of their armored forces. They continued either to advance timidly when there was scarcely any resistance left, or else they carried out deep, isolated tank thrusts which the infantry was unable to follow and which, consequently, could not lead to permanent success. Russian armored forces always incurred severe losses wherever they encountered German armor still organized in units of any appreciable strength. Thus, as late as April 1945, the battle-weary German 6th Panzer Division succeeded, in what was probably the last tank battle, in repulsing vastly superior Russian tank forces in the plains of the lower March River, and in knocking out 80 tanks.

If the Russian tank forces with their vastly superior numbers had had proper leadership, the Russians would have been able to bring about the end of the war at a much earlier stage.

Chapter 10

Defense

I. General

There are two conspicuous characteristics of the Russian soldier, both of which are inherent in the Russian people, both of which were in evidence during the course of the two world wars: stubbornness and tenacity in the defense, inflexibility and little adaptability in the attack.

During World War I the power of resistence of the Russian soldier was gradually paralyzed because of a lack of heavy weapons, the great inferiority of Russian artillery, the almost complete lack of aircraft, and because the morale of the Russian Army was undermined more and more as time went on. As early as 1915 Russian units left their trenches at the onset of German attacks, came toward the attackers, and surrendered. Such incidents, however, were not characteristic of the Russian soldier, but rather an indication of conditions prevailing at that time. Elite units, such as Guards, and ably commanded troops defended themselves with extreme doggedness even in World War I.

During the first phase of World War II there were also numerous examples of Russian soldiers showing but little power of resistance, throwing away rifles, and surrendering or deserting by the thousands. This, however, occurred only in great battles of encirclement where the Russian soldier became demoralized by Stuka attacks and heavy, concentrated artillery fire and realized the impossibility of continuing the battle successfully. In general, in this war, firm leadership, good equipment, emphasis on patriotism, and the fear—stimulated by propaganda—of falling into enemy hands, resulted in a tenacity of defense which made German attacks without tanks, despite artillery and air support, extremely costly or even futile. This power of the Russian soldier to resist increased during the course of the war in a direct ratio to the decrease of the German power to attack. The numerical superiority of the Russians became more and more crushing, their equipment continued to improve as compared with German equipment, and continuous military successes lifted their spirit and confidence.

When on the defensive the Russian disappeared into the earth with amazing speed. Within the shortest imaginable time he constructed a system of field fortifications with numerous earthworks of all kinds.

Laying mines of various types and stringing barbed wire also took very little time. The troops were deployed in depth; the Red commanders frequently changed the strength of forces occupying a position (even to the extent of varying the strength of the day and night shifts), and immediately prepared a careful plan of fire for all weapons. Except during great battles, defensive fire was maintained chiefly by the infantry heavy weapons, particularly by mortars which the Russians employed with considerable flexibility and in great numbers. In attacks against Russian positions, it was usually quite a while before effective defense fire of the Russian artillery began. But even after a long interval the Russian artillery was unable to direct its fire precisely and flexibly. The emphasis of the defense was on the infantry weapons, including antitank and self-propelled assault guns, and on extensive mine laying.

From the German point of view, the defensive power of the Russian troops is none too good if the attacker builds up a strong superiority in air and ground weapons and provides for sufficient depth of the attacking force. Above all, the attacker must start determinedly and must know how to exploit each success quickly and flexibly in order to achieve freedom of movement as rapidly as possible. The Germans believe that a resourceful commander who knows how to take advantage of the initiative will certainly be successful, because the Russian command does not possess the necessary speed of action in unexpected situations.

In World War II the Russian was a master of the defense. He attained excellent results not only in construction of positions, but also in camouflage and in the construction of dummy installations. By unscrupulous use especially of the civilian population (including women and children), he created well-developed zones in depth. In open terrain he dug wide and deep antitank ditches, often many miles long. Mine fields, wire obstacles, entanglements, and other obstacles were immediately set up everywhere. If, because of the nature of terrain, he expected tank attacks, the enemy developed points of main effort. He was very adept at using villages as strong points. Wherever he could, he set up flanking weapons. He conducted his infantry defense in a mobile manner, but within his defense trenches. He made considerable use of roving guns. He accomplished surprise fires mostly with heavy mortars and numerous multibarreled rocket projectors. He made little use of methodical artillery harassing fire. Upon the loss of parts of a position, reserves for a counterthrust were always quickly at hand. Counterthrusts and counterattacks were in most instances supported by tanks. The Russian did not often conduct counterattacks that were based on a preconceived plan since, from the standpoint of leadership, they were too difficult for him. From

1943 on, he strengthened his defense by mass employment of antiaircraft guns and flame throwers in so-called antitank gun fronts, which were superbly camouflaged and dangerous for tank attacks. Daytime observation was made difficult, because he showed almost no movement. In general, one might say that the Russian undertook his defense in open terrain, as well as wooded areas, according to principles rigidly drilled into the soldiers, and that he showed little imagination in developing new methods of battle. He relied, in attack as well as in defense, on reckless employment of manpower.

Another peculiarity to be mentioned is the fact that the Russian proved himself to be very well disciplined in opening fire. He waited calmly until the Germans had approached to a favorable range, and then he opened surprise fire. German combat reconnaissance always had to be on the lookout to avoid unnecessary losses. The Russian sniper battalions particularly excelled in fire discipline.

In 1941 the Russians also supported defense action in the Parpach position in the Crimea very adroitly and effectively from the sea by artillery flanking fire.

II. Use of Mines

The Russian made extensive use of mines. As a rule, a protective mine belt was to be found about 8 to 10 yards in front of the most forward trench. Terrain particularly favorable for an enemy approach likewise was heavily mined. The Russian preferred to employ wooden box mines which could not be detected by the standard mine detectors. In the depth of the battle position, mines were laid in unexpected places. In favorable terrain, antitank mines were numerous.

Difficulties in transporting Teller mines because of the lack of transport space were solved in a very primitive manner. When marching up to the front as relief, every man had to carry two antitank mines. At the front, these mines were laid by the engineers according to a diagrammed mine plan.

In 1944–45 the Russians, while on the move, also scattered mines around points of main effort in order to block tank attacks. In the southern Ukraine, following a successful tank thrust, the Russians immediately protected the terrain they had gained with a belt of antitank mines blocking all roads and approaches. On 1 day alone, 20,000 such mines were laid. German counterattacks ground to a halt and collapsed in mine fields of that type.

The Russian cleared mines in front of German obstacles during the night, and used them for his own purposes. Later on, the Germans laid mine fields only behind their own front, at points at which a tank break-through or an enemy offensive was to be expected.

When the Russian intended to give up a previously defended zone, he used many tricks. For example, he attached demolition charges with push-pull igniters to abandoned field kitchens, weapons, corpses, and tombstones; he connected explosive charges to doors, windows, or stoves in the winter; he installed pressure mines under stairs and floors, and booby-trapped abandoned trucks and other equipment.

When the Germans took Kiev and Vyborg in 1941, and Sevastopol in 1942, the Russian used remote (radio) control mines to blow up entire blocks of houses as soon as anyone entered. This type of remote control ignition seemed to be the most effective solution to the problem. The apparatus consisted of three parts, each of which was not much larger than a full brief case. It was easy to move, could easily be built into a hidden spot, and was at first hard to find. Built-in safety devices prevented an accidental or untimely detonation. The ignition apparatus included a clock, which ran only on a certain signal for a certain length of time, and permitted detonation only during certain minutes. There was, furthermore, a code which, when given at a certain speed, was the sole means for setting off the mines. The ticking of the clocks, however, could be heard with sensitive sound detectors and often led to discovery.

During the autumn offensive in 1941 against Moscow, the Russian employed so-called mine dogs for destroying German tanks. In the manner of pack animals, medium-sized dogs carried demolition charges which were connected to a spindle fastened to the dog's back. The dogs were trained to hide under approaching tanks. In so doing, the animal inadvertently brought the upright spindle which was about 6 inches long into contact with the belly of the tank and set off the charge.

News of this insidious improvisation caused some alarm in the German panzer units and made them fire at all approaching dogs on sight. So far, there is no evidence of any case where a German tank was destroyed by a mine dog. On the other hand, it was reported that several mine dogs fleeing from the fire of German tanks sought protection underneath Russian tanks which promptly blew up. One thing is certain: the specter of the mine dogs ceased just as abruptly as it had begun.

III. Conduct of Battle

The Russian defended every inch of his soil with incredible tenacity. At the beginning of the war he was conversant only with long defense lines, which he strengthened by employing an amazing number of personnel for digging trenches. Artillery confined its activities to minor fire concentrations. However, it proved to be more mobile than the infantry and employed many roving guns which sometimes became very annoying to the Germans, since they fired only a short time from one position and then reappeared elsewhere. The Russian required a long period of time to bring effective fire to bear against an attack already in progress.

In 1943 new methods of tank and antitank warfare were introduced, though the methods of the infantry remained the same. Heavy machine guns and dug-in tanks frequently were encountered deep in the battle position. The latter were particularly dangerous because they were well armored and difficult to hit. In sectors in which the terrain was passable for tanks, antitank gun fronts would be set up in nearly all instances. They were developed to extraordinary strength and foiled many a German armored attack. They were also used against infantry, in which case they fired high-explosive shells.

In most cases, tank counterattacks without infantry support were certain to fall victim to German antitank defense. Infantry counterattacks were infrequent and generally too late. For the rest, no important changes took place in Russian defense methods up to the end of the war. The multibarrelled rocket projectors, which the Russians employed more and more, certainly were very effective psychologically but their physical effect was much less impressive. The most dangerous weapons continued to be the medium and heavy mortars, particularly after the heavy mortars were concentrated in battalions in 1944.

Areas of great importance were surrounded with heavy fortifications. For instance, the German armored units before Leningrad encountered fortification systems up to 6 miles in depth, including innumerable earth and concrete bunkers with built-in guns and other heavy weapons. There were even concrete pillboxes with disappearing armored cupolas for artillery and machine guns. They were constructed in the rear area from standard concrete forms, assembled at the front, and equipped with the armored cupolas. The cupolas were raised and lowered by wooden levers, which had to be operated manually by the pillbox crew. A speedy elimination of these concrete pillboxes with the means available in mobile warfare was difficult.

The forward edge of such a defense system was generally situated behind an antitank ditch several miles long and up to 20 feet wide and 12 feet deep. Embedded in the rear wall of this ditch were dugouts housing the riflemen with their defense weapons. A second and third antitank ditch frequently would be located in the depth of the system, and connected by a cross ditch so as to prevent enemy tanks that had penetrated the position from rolling it up. A machine-gun or antitank bunker in every bend of the antitank ditch afforded flank protection. It was not unusual to encounter dammed-up watercourses close to the fortified position. They were up to a hundred yards wide and several yards deep, and presented an obstacle difficult to overcome. The Russians eliminated all favorable approaches to their front (forests, underbrush, tall grain fields, etc.) by laying extensive mine fields.

Outposts were located ahead of the fortified position wherever possible. Such outposts always had engineers attached whose mission it was to block routes of approach with mines or other obstacles.

The area of Krasnogvardeysk, south of Leningrad (Map 8), had been developed according to the above-mentioned principles into an outlying fortress. During early September 1941, it presented great difficulties to the advance of several German corps. Krasnogvardeysk blocked all highways and railroads leading to Leningrad from the south, thus constituting a main bulwark of Russian resistance. The Russians defended it persistently. Repeated attacks by several infantry divisions were repulsed. Only in the course of a general attack on the Leningrad Line, and after bitter pillbox fighting in the area immediately surrounding Krasnogvardeysk, was that town finally taken from the rear by a carefully prepared surprise break-through to the west of it. This typical example of the Russian method of defending a methodically fortified zone, and its capture by an adroitly led attack, will be discussed in detail in the following paragraphs.

The defense system of Krasnogvardeysk had been prepared long in advance and consisted of an outer belt of concrete and earth bunkers, with numerous intermediate installations which were interconnected by trench systems that could easily be defended. There were tank-proof watercourses or swamps almost everywhere in front of the outer defense belt. Where this natural protection was lacking, wide antitank ditches had been dug.

At a distance of 1,000 to 3,000 yards behind the outer defense belt there was an inner one consisting of a heavily fortified position encircling the periphery of the town. Just north of the town ran the continuous Leningrad Line, with which the defense system of Krasnogvardeysk was integrated. It constituted, simultaneously, the rear protection of the town and the covering position in case the town should have to be evacuated. Beyond the open, elevated terrain immediately west of Krasnogvardeysk lay an extensive forest zone. Within that zone, a few hundred yards from its eastern edge, ran the western front of the outer defense belt. At that point it consisted of wood and earth bunkers, trenches, and strong points—all approaches to which were barricaded by extensive mine fields, abatis, and multiple rows of barbed wire. Located from 2 to 3 miles farther west were mobile security detachments. Attached to these were engineer units used to lay scattered mines.

The cornerstone of this position was the heavily fortified and mined village of Salyzy, located at the southern end of the forest zone. It covered a road leading to Krasnogvardeysk from the west, and another one which branched off the former to the north within Salyzy. The later road served as supply route for all the troops situated west of it in the forest position. It crossed the dammed-up Ishora River via a bridge located in front of the Leningrad Line, traversing the line in a northwesterly direction. At that point the line consisted of four trench systems, one behind the other, with numerous machine-gun, antitank-gun, and artillery bunkers.

The German 6th Panzer Division, advancing on Krasnogvardeysk from the west via Salyzy, had the mission of breaking through the Leningrad Line in the above-described area, and attacking Krasnogvardeysk from the rear. Following a plan of attack based on precise aerial photos, the division decided to push with concentrated force through the outer defense belt at Salyzy, to follow through with a northward thrust and break through the Leningrad Line, and then to roll up the later to the east. The main body of the division attacked on the road and along the edge of the forest running parallel to it, took the antitank ditch after a brief engagement, and, during the noon hours, also captured the village of Salyzy after having stormed a large number of bunkers. A bunker at the edge of the forest continued to offer resistance until late afternoon.

Immediately after breaking into the village, the armored elements of the division, supported by an artillery battalion, advanced through the rear of the enemy-occupied forest position against the Leningrad Line. Under cover of tank fire, the engineers took the undamaged bridge in a *coup de main* and removed prepared demolition charges. About 4 miles north of Salyzy, panzer grenadiers (armored infantry) following them penetrated the antitank ditch, which began at the bridge and ran at a right angle to the front, and formed a bridgehead. During the evening the main body of the division cleared the surrounding forests of Red forces and, with a front turned 90°, assembled in the woods for a northward thrust to join the forces at the bridgehead, which meanwhile had been cut off by the enemy. On the evening of the same day a German panzer grenadier battalion succeeded in breaking through the inner defense ring located east of Salyzy behind a river arm, in the area of the neighboring SS Police Division, which had been stalled along its entire front. The bridgehead thus established by the battalion opened a gateway to Krasnogvardeysk for the SS Police Division.

On the second day of attack the bulk of the 6th Panzer Division advanced along the road to the armored units in the northern bridgehead. Some of its elements mopped up enemy forces on the plateau west of Krasnogvardeysk, while others rolled up the Russian forest position which was pinning down a unit that had been detached as a flank guard on the previous day. During the forenoon the entire attack area south of the Leningrad Line was cleared of the enemy. Along the northern wing of the position on the forest edge alone, 40,000 Russian mines were picked up and disarmed. Then, battalion after battalion was pushed through the bridgehead into the 2-mile-long antitank ditch which ran up to a forest area. These battalions were pushed so far to the north that the four parallel defense systems of the Leningrad Line could be rolled up simultaneously from the flank by one battalion each. A desperate enemy attempt to repel the advance of the battalions and tanks by a cavalry attack was easily foiled. The antitank ditch, 4 yards wide and deep, had made it possible to change the attack front of the entire division again by 90° at one stroke. Under cover of flank and rear protection, bunker after bunker and strong point after strong point was now blasted by Stukas, medium artillery, antitank and antiaircraft guns, and captured; step by step the trenches and nests of resistance were cleaned out. All the German artillery was still in the old front south of the Leningrad Line, and its fire thus formed a complete flanking curtain in front of the attacking battalions.

The railroad running through the attack area was reached on the second day of battle, and the Krasnogvardeysk–Leningrad highway on the third day. There, the German forces took a group of artillery pillboxes equipped with disappearing armored cupolas. At that point the division stood directly in the rear of the town. The enemy, forced to retreat hurriedly, had only one side road available for a withdrawal, and that road lay under the effective artillery fire of the panzer division. With serious losses, the Russian divisions poured back over this road and the adjoining terrain. The attempt of the motorized medium artillery, the first of the Red forces to disengage, to escape on the wide asphalt road via Pushkin, failed. The road was already blocked by German armor. All the Russian artillery, as well as all other motor vehicles, was set afire by German armor when it attempted to break through at this point. During the following night the Russians, although badly mauled, managed to evacuate the town and escape. They then reestablished themselves with strong rear guards on high terrain between Krasnogvardeysk and Pushkin.

The next day, pursuing German infantry divisions bogged down before the heavily fortified positions. Here the enemy had employed the most modern system of field fortifications ever encountered on the Eastern Front. All of the fortification installations were underground. The defense was carried out in subterranean passages which were established along terrain steps and were equipped with well-camouflaged embrasures. The heavy weapons likewise were in subterranean emplacements which were invisible from the outside. There were also subterranean rooms quartering 10 to 20 men each, ammunition dumps, and medical and supply installations. All installations were interconnected by underground communication passages. The entrances were situated several hundred yards farther to the rear, well camouflaged by shrubbery and groups of trees, and protected by open squad trenches and several standard bunkers which could only be recognized from nearby. Neither the best ground nor air reconnaissance could spot this fortification system even at close range. Not even after its guns had opened fire could it be located, as a result of which it proved very difficult to neutralize. All frontal assaults of the infantry were unsuccessful.

Not until 2 days later was it possible to clarify the situation and to capture the position. By that time the 6th Panzer Division, committed as an encircling force maneuvering via Posyolok Taytsy, was pivoting into the rear of the Russian fortifications. An odd coincidence played into the hands of the division. The previous evening strong reconnaissance patrols had advanced into the high terrain. Suddenly encountering the rearmost outlying bunker of the position, the patrols took it by storm without orders. Among the captured garrison was a Russian military engineer, the builder of this fortification system. With him, the plans of all the installations fell into German hands, and it was easy to plan the attack for the next day.

However, the attack of the lead-off panzer grenadier regiment had hardly begun when a new difficulty arose. The Russian had recognized the danger to his frontally impregnable position, and launched an attack from Pushkin against the rear of the 6th Panzer Division. A long column of tanks, the end of which could not even be surmized in the dust, rolled against the German rear guard unit. The first of the Russian tanks had already passed a narrow strip between swamps and turned against the defended elevations. However, except for one 88-mm. Flak battery and the antitank guns of the rear guard panzer grenadier battalion, the German division had at that moment only one panzer battalion with light Czech tanks available for its defense. The heavy Flak guns were already thundering. Flames from tanks that had sustained hits rose straight toward the sky. The vanguard of Red tanks consisting of 54 KV1's spread out, but kept moving ahead. Suddenly it was attacked and destroyed from very close range by a hail of fire from the tank destroyer battalion which had just arrived with 27 heavy antitank guns. Fourteen columns of black smoke announced to the main body of the enemy the destruction of his vanguard. Thereupon the main body suddenly stopped and no longer dared to pass the swamp narrow. Rear elements fanned out and disappeared into the adjoining terrain. Heavy tank fighting indicated that the German panzer regiment, which had been summoned by radio, had gone into action. Soon the din increased. The panzer regiments of the neighboring 1st and 8th Panzer Divisions which had also been summoned attacked the flank and rear of the Red forces. The Russian realized his precarious situation and felt himself no longer equal to the task. His losses and his retrograde movements bore evidence to that fact. Even the heavy tanks, only 14 of which had been reached and destroyed by the heavy Flak guns, turned and retreated. The enemy had avoided a showdown. The threat to the rear of the panzer division had thus been eliminated.

In the meantime, however, the German panzer grenadier attack, supported by a panzer battalion, continued according to plan. In heavy fighting, the bunkers and squad trenches which protected the enemy's rear were taken one by one, and the entrances to the subterranean defense system reached. During the fighting for the first entrance, the crew resisted from an inner compartment with fire and hand grenades. In this action three Russian medical corps women in uniform, who defended the entrance with hand grenades, were killed. When their bodies were removed, several hand grenades were found on them.

Mopping up the subterranean passages was time-consuming and difficult. It had to be carried out by specially trained shock troops with hand grenades and machine pistols. German attempts to clear out the strong bunker crews led to bitter underground hand-to-hand fighting with heavy losses on both sides. The enemy defended himself to the utmost. The attack stalled. Only after engineer demolition teams had succeeded in determining the location of the subterranean bunkers by noting the sparser growth of grass above them, could these bunkers be blown up by heavy demolition charges from above, and taken. But the closer the shock troops came to the front position of the enemy's defense system, the more serious became the losses. The engineer demolition teams and all the other units were advancing above the Russian defense system, into the heavy artillery fire supporting the frontally attacking German infantry divisions of the neighboring army. Only when roundabout telephone communications had been established, and the devastating fire ordered to cease, was it possible to take the entire subterranean defense system. A junction was then effected with the infantry on the other side. Subsequently, German forces also occupied Pushkin.

With that, the most tenacious Russian defensive battles of 1941, between Krasnogvardeysk and Leningrad, came to an end. Only the flexible leadership of battle-tested armored forces, attacking with elan, made it possible to overcome the defense zones which had been set up in an all-out effort of the latest Russian defense technique. Within a week the German 6th Panzer Division had had to break through and roll up 12 positions, repel several counterattacks, and take more than 300 heavily fortified bunkers.

Equally instructive was the Russian conduct of battle in the defense against the German pincers attack (Operation ZITADELLE) on Kursk in July 1943. The exhaustion on both sides after the preceding long winter battles led, at this sector of the front, to a pause of 3 months which both opponents used to replenish their forces and to prepare for Operation ZITADELLE. The Russians expected the attack precisely at the location and in the manner in which it was undertaken, and prepared their defense accordingly.

Behind the most endangered sectors, opposite Byelgorod and Orel, they constructed defense systems of hitherto unknown depth, and strengthened them with all kinds of obstacles. To be prepared against surprise armored thrusts, all points susceptible to penetration were safeguarded up to a depth of 30 miles by fully manned antitank gun fronts, antitank ditches, mine fields, and tanks in emplacements, in such numbers and strength that to overcome them would have called for great sacrifices and much time. Behind the pressure points north of Byelgorod and south of Kursk, sufficient local forces stood ready everywhere. Noteworthy were the numerous alternate firing positions, and the fact that the bulk of the numerous Russian artillery pieces were kept as far to the rear as their maximum range allowed, so as to escape counterbattery fire from German heavy howitzer batteries and to be able, in case of reverses, to support the infantry as long as possible. The Russian batteries preferred firing positions in forests, or in orchards adjacent to inhabited localities. For mobile operations, the Russians very adroitly employed multibarrelled rocket projectors. Strong strategic reserves were assembled farther east, in the region of the Oskol River, in such a manner that after the attacking German divisions had exhausted themselves in the above-mentioned defense system, the reserves could launch a counterattack, or, at worst, contain an enemy break-through. In the bulge extending far to the west, however, the enemy had stationed only weak and inferior forces, which were not backed by any deep defense system. During the long waiting period each side learned about the other's situation and intentions down to the last details. The Russians, for instance, broadcast to the German lines by loud-speaker the secret day and hour of attack well in advance, and in the same manner announced two postponements of the offensive. Nevertheless, the German attack was carried out at the precise point at which the Russians expected it. As anticipated, it did not develop into a dynamic offensive, but became a slow wrestling match with an enemy firmly clinging to a maze of trenches and bunkers—an enemy who, unshaken by preparatory fire, offered dogged resistance. Many positions could only be taken after prolonged hand grenade duels. The Russians employed stronger tank forces only against what they guessed to be the weakest point in the German attack wedge—the flank of XI Infantry Corps which attacked on the right wing. Every one of these counterattacks was repulsed.

On the very first day of the attack, 5 July 1943, several German divisions each sustained losses up to 1,000 dead and wounded. The German armor, too, suffered substantial losses each day from the strong antitank defenses and mine fields. This, as well as the divergent directions of thrust of the various corps, visibly diminished the momentum of the German attack. When, after about 2 weeks of bloody fighting, there was no longer any hope of reaching the desired objectives, and when German forces even began to meet reverses in the Orel area, the attack was called off, and previous territorial gains were relinquished. By excellent organization of defenses and adroit conduct of battle, the Russians had brought about the collapse of the German offensive. Shortly thereafter they launched a counteroffensive with fresh reserves and effected a major break-through.

Conditions were entirely different for the Russians in their defense against the German relief thrust on Stalingrad in December 1942. (Map 9) Here, there existed only temporary field positions, and the defense had to be conducted in a mobile manner. At first the Russians pushed a cavalry corps, strengthened by armor and camel troops, forward along the Don to the Kurmoyarskiy Aksay River for reconnaissance and for screening the movements of their infantry and tank forces assembling in the rear.

The German 6th Panzer Division was 10 to 20 percent overstrength and had to conduct the main thrust. When its leading elements arrived, the vanguard of the Red calvary corps was just moving into Kotelnikovo (about 26 November 1942). It was driven back, and the assembly of German forces continued. The attempt to take Kotelnikovo in an assault by the entire cavalry corps on 5 and 6 December 1942, ended in a smashing defeat of the corps at Pokhlebin. Meanwhile, the enemy cautiously advanced two rather weak infantry divisions along both sides of the railroad onto the elevations north of the city, and pushed back several outposts. After the bitter experience of Pokhlebin, however, he did not dare attack Kotelnikovo again. He assembled his main force, the Third Tank Army and additional infantry forces, between the Aksay River and the Mishkova River sector. His entire defense forces were drawn up in three echelons, one behind the other, 20 miles in width and 45 miles in depth. The impression was gained by the Germans that the enemy would move up under the protection of his advance infantry and cavalry divisions, and then, with his entire tank army, attack the 6th Panzer Division which was marching up alone, in order to destroy it in the wide forefield of Stalingrad before it reached the city. The move, however, did not materialize. On that occasion the Russians either missed a chance, or else did as yet not feel strong enough to attack the division, which was equipped with 200 tanks and self-propelled assault guns as well as a large number of antitank weapons. Neither did he act to save his reinforced cavalry corps from destruction on 5 and 6 December, and also looked on idly on 12 December while the beginning of the relief thrust of the German 6th Panzer Division rolled over his advance infantry divisions and scattered them. The northernmost of the two divisions here lost its entire artillery. The weak remnants of the Russian cavalry corps were also caught on the fringes of the mighty assault and so badly mauled that they played no further part in the course of the offensive. Thus the 6th Panzer Division, without protection on its northern flank, was able to cross the Aksay River as early as the third day. Its southern wing was protected by the 23d Panzer Division (in regimental strength with 15 to 20 tanks), which followed in echelons.

The crossing of the Aksay River met only weak resistance from advance elements of a Russian mechanized corps, which was soon overcome. In an immediate follow-up thrust by all German armored units, Verkhniy-Kumskiy, the key point of the assembly area of the Russian Third Tank Army, was taken. Not till then were the enemy tanks stirred to action, but now they displayed very spirited activity. Speed was imperative. Therefore, the Russian commander was compelled to radio all his orders and reports in the clear. Because the Russians were forced to put their cards on the table, the German forces, although numerically far inferior, were able during the ensuing several days of bitter tank fighting to attack Russian elements in lightning moves and beat them decisively before they could receive help. In the melee that followed, the Russians occasionally succeeded in concentrating greatly superior forces which threatened to become dangerous to the division. The German armored forces then immediately withdrew, only to attack the Reds from the rear again the moment an opportunity presented itself.

Both sides made large-scale shifts under cover of darkness. By lightning-like feints and changes of direction, it was repeatedly possible for the Germans to attack strong Red tank concentrations simultaneously from all sides in the larger hollows of the hilly terrain, and to destroy them to the last tank. In this manner a number of so-called tank cemeteries originated, where from 50 to 80 knocked-out tanks, mostly T34's, stood in clusters within a small area. German bomber wings repeatedly bombed them by mistake. Neither German nor

Russian aircraft could take any part in the seesaw tank battles, since the opposing tanks were frequently so intermingled that they could not be differentiated. Although air activity on both sides was very lively, it was forced to limit itself to attacking motor pools and supply lines. The air arm was of no decisive importance.

While the tank battle north of the Aksay River was still in progress, Russian tank and motorized brigades crossed the river in a southerly direction and attempted to cut off the bridge crossing, which was strongly held on both sides of the river. This had to be avoided under all circumstances, but without depleting the German armored forces then engaged in crucial battles. Soon the bridgehead was surrounded. Although more than a dozen Russian tanks were knocked out, just as many surviving tanks overran the entrenched German infantry and penetrated to a rather large village located in the center of the bridgehead and defended by the German 57th Armored Engineer Battalion. Not a single engineer or rifleman deserted his post. Each man became a tank buster. Just as fast as Russian tanks entered the village, they burst into flames. Not one escaped. Three times the Russian repeated this assault, and three times he was repulsed. Then a reinforced infantry regiment attacked him from the rear, scattered the entire force, and knocked out 14 tanks. That opened the route of advance again and assured the free flow of supplies across the bridge.

When strong Russian motorized infantry with numerous antitank weapons entered the battle at Verkhniy-Kumskiy, the German armored infantry forces were tied down supporting their neighbor and other forces engaged at the Aksay River, leaving the German armored units so severely restricted in their freedom of movement that they had to be withdrawn to the Aksay River. The Russian, however, had suffered such heavy tank losses that he did not dare risk the rest of his tanks in a pursuit. He contented himself with defending a long ridge south of Verkhniy-Kumskiy. The premature attempt, ordered by a higher command, to roll up this ridge position from the flank with the combined armored elements of the 6th and 23d Panzer Divisions failed. Its failure was caused by lack of sufficient infantry to silence the numerous antitank guns and rifles which were entrenched in deep antitank pits and well camouflaged by high steppe grass. Although it was perfectly possible to roll from one end of the ridge to othe other, the Russian motorized riflemen popped up again afterwards like jack-in-the-boxes and, with their numerous antitank rifles, knocked out many an armored vehicle. The combined German armored force suffered considerable losses and had to be recalled in the evening without having accomplished its mission.

Not until 2 days later did a planned attack of the entire 6th Panzer Division succeed in taking the position and cleaning it out. In the subsequent night attack the German armored infantry recaptured the stubbornly defended village of Verkniy-Kumskiy, destroyed a number of emplaced tanks, numerous antitank guns, and over 100 antitank rifles. At dawn of the following day the elevated position north of the village was taken in cooperation with the newly arrived 17th Panzer Division which had only the combat strength of a reinforced battalion. The 11th Panzer Regiment, which up to this time had been held in reserve, was employed in the pursuit and inflicted heavy losses upon the Russian who was retreating through a single defile.

In the midst of the pursuit, however, the entire 6th Panzer Division had to turn to the east to support the neighboring 23d Panzer Division on the right since it was being pushed back beyond the Aksay River by a newly arrived Russian rifle corps. The further pursuit toward the north had to be left to the weak 17th Panzer Division, which lacked sufficient driving force to destroy the beaten Russian forces.

The turning of the 6th Panzer Division against the rear of the new enemy had decisively changed the situation in the 23d Panzer Division. The Russian corps immediately broke off its attack and hastily retreated eastward in order to escape the deadly blow that would have been dealt it very soon, had it remained. The Red tanks and antitank gun fronts thrown against the German 6th Panzer Division had been scattered before, and German armor was about to cut off the Russian escape route. At this critical moment, too, the Russian corps commander radioed his urgent orders in the clear.

But the objective of the German panzer division was Stalingrad, not the pursuit of a corps in a different direction. As soon as the 23d Panzer Division, relieved of enemy pressure, could again advance, pursuit of the corps was halted. The 6th Panzer Division then turned north and, after hard fighting, reached the Mishkova River sector at Bolshaya Vasilyevka. At that point the Stalingrad garrison was supposed to make contact with the division. Two bridgeheads were quickly formed, the village taken, and the entire division concentrated in a small area for mobile defense. It had already covered two-thirds of the distance, and stood 30 miles from Stalingrad; the flash of signal rockets from the city could be observed at night. It remains a puzzle why the German Sixth Army (Field Marshal Paulus) did not break out at that time (20 December).

In forced marches the Russian brought up additional strong forces from the Stalingrad front and the Volga in order to support the beaten Third Tank Army and throw back the German forces. Since he no longer had sufficient tank forces available for this purpose, he hoped to overwhelm and destroy them with the newly formed infantry main-attack army. The Red riflemen surged forward in multitudes never before encountered. Attack wave followed attack wave without regard for losses. Each was annihilated by a terrific hail of fire without gaining so much as a foot of ground. Therefore, the Russians went around the two flanks of the German division in order to encircle it. In the course of this maneuver they came between the German artillery position and the panzer regiment. Firing from all barrels, 150 tanks and self-propelled assault guns attacked the Russian masses from the rear when they tried to escape the fire from the artillery. In their desperate situation many Russians threw down their weapons and surrendered. Succeeding elements flowed back; Red forces which had penetrated into the village were driven out again by a counterthrust of the German infantry, and Russian tanks which had broken through were knocked out. The Russian mass assault had collapsed.

On 22 December the German 6th Panzer Division had regained its freedom of movement. By a further forward thrust of 18 miles on 24 December, the division was to help the encircled Sixth Army in breaking out of Stalingrad. That operation, however, never materialized, because the division suddenly had to be withdrawn on 23 December and transferred to the area north of the lower Don (Morosovskaya) to bolster the collapsed Chir front. This move definitely sealed the doom of the German forces at Stalingrad. The remaining two weak panzer divisions, the 17th and 23d, were not even sufficient to make a stand against the Russian forces, let alone repulse them. But also the enemy was so weakened by his losses, which included more than 400 tanks, that he was unable to make a quick thrust against Rostov, an action which would have cut off the entire Caucasus front.

Here is another example that confirms very emphatically the characteristic fighting method of the Russians: not great achievements by small units with clever leadership, but by sacrifices of masses. Only when the Russians attacked with a tenfold to twentyfold superiority could they achieve temporary successes. In the assault, however, the individual performance of the German soldier triumphed over the masses.

The Russian high command had assigned more than sufficient forces to prevent the relief (five corps and one main-attack army against one German panzer corps). The Russian corps were organized very effectively, but were poorly led. If properly led, these superior forces would have sufficed to defeat the weak German relief force before it could launch its attack. At the very latest, the combined Russian corps should have attacked and beaten this German corps which was making an unsupported forward thrust, when it crossed the Aksay River. But the commitment in detail of the Russian forces enabled the German units to attack the individual Russian corps by surprise and defeat them one after another. During the last days, the Russians hastily formed a new main-attack army and threw it against the German 6th Panzer Division in the Mishkova River sector in order to halt its further advance. This army also sustained heavy losses, and would have been unable to prevent a further advance of the German armor. It had no effective tank support because the pivotal element in the whole struggle, the Russian Third Tank Army, had already been beaten. The Third Tank Army was the most dangerous opponent on the route to Stalingrad. It had more than twice the number of tanks the Germans had, and far superior antitank weapons. Its motorized troops were well trained and fought with exemplary valor. But neither numerical superiority nor valor could make up for the mistakes of the intermediate and lower commands. Only the repercussions of the great successes which the Soviet Supreme Command was able to achieve on the Chir and Don fronts caused the German relief thrust to fail.

An indirect but very significant role was also played by the Allied invasion of North Africa. The German armored forces stationed in France (the 6th, 7th, and 10th Panzer Divisions and the 1st, 2d, and 3d SS Panzer Divisions) were among the crack troops of the German Army. They had been brought up to full effectiveness, but were not transferred as a whole to the East for fear of an Allied landing in southern France. Finally, after several days' hesitation, only the 6th Panzer Division was transferred to the sector south of Stalingrad. The 7th Panzer Division followed at a later date, but arrived too late for the relief thrust. Its timely arrival would have been sufficient to carry the thrust through to its objective.

Chapter 11

Retreat and Delaying Engagements

Observations of Russian fighting methods in retreat could be made particularly during the first years of the war. When the Russians had been defeated on a broad front, they reestablished their lines only after they had retreated a considerable distance. They marched very quickly, even when retreating in extremely large numbers. Precisely at such times it was important to pursue them energetically, and to give them no opportunity for renewed resistance. The German conduct of delaying action, with leapfrog commitment of forces in successive positions, was apparently not known to them; at least, this method of fighting, requiring great mobility and competent leadership, was not used by the Russians. The Russians always sought only simple and complete solutions. When they decided to withdraw, they did so in one jump, and then immediately began an active defense again. When German armored forces which had broken through chased them off the roads, the Russians disappeared into the terrain with remarkable skill. In retreating, retiring from sight, and rapidly reassembling, the Russians were past masters. Even large forces quickly covered long distances over terrain without roads or paths. In 1941 certain Russian rifle regiments which had been thrown back by German armor crossing the border at Tauroggen, again opposed the same panzer division south of Leningrad, after a march of 500 miles.

Even if time was of the essence, the Russians succeeded in carrying off large numbers of cattle, as well as a substantial amount of equipment and supplies. They shot thousands of undesirable persons in the Baltic countries before the retreat, and took other tens of thousands with them. In retreating they did not hesitate to burn to the ground the cities and towns of their own native land, if it seemed that any advantage was to be gained (scorched-earth policy). Thus, in the retreat of 1941, they almost completely destroyed the cities of Vitebsk, Smolensk, and many others, leaving nothing of value to fall into the hands of the Germans and delaying their advance. All that remained for the Germans of the Russian collective farms, state farms, machine tractor stations, and manufacturing plants of all kinds, were ashes and ruins. For that reason it even became difficult in some sectors to quarter larger headquarters organizations to assure their ability to function.

Russian tactics of stopping or slowing down the German offensive by uncoordinated counterattacks have already been treated in the preceding chapter.

Chapter 12

Combat Under Unusual Conditions

The Russians systematically exploited all difficulties which their country presented to the enemy. In villages, woods, and marshes, and in fog, rain, snow, and storm, the Russians combined the tricks of nature with their own innate cunning in order to do the greatest possible harm to the enemy.

I. Fighting in Towns and Villages

The Russians were very adept at preparing inhabited places for defense. In a short time, a village would be converted into a little fortress. Wooden houses had well-camouflaged gun ports almost flush with the floor, their interiors were reinforced with sandbags or earth, observation slots were put into roofs, and bunkers built into floors and connected with adjacent houses or outside defenses by narrow trenches. Although almost all inhabited places were crammed with troops, they seemed deserted to German reconaissance, since even water and food details were allowed to leave their shelters only after dark. The Russians blocked approach routes with well-camouflaged antitank guns or dug-in tanks. Wrecks of knocked-out tanks were specially favored for use as observation posts and as emplacements for heavy infantry weapons, and bunkers for living quarters were dug under them. It was Russian practice to allow the enemy to draw near, and then to fire at him unexpectedly. In order to prevent heavy losses of personnel and tanks, the Germans had to cover the outskirts of inhabited places with artillery, tanks, or heavy weapons during the approach of their troops. Fires resulted frequently, and in many instances consumed the whole village. When the front line neared a village, the inhabitants carried their possessions into outlying woods or bunkers for safekeeping. They did not take part in the fighting of the regular troops, but served as auxiliaries, building earthworks and passing on information. The Russian practice of raiding inhabited localities during mobile warfare, or converting them into strong points for defensive purposes, was responsible for the destruction of numerous populated places during combat.

Since the defenses on the outskirts of a locality were quickly eliminated by the above-mentioned German countermeasures, the Russians later led their main line of resistance right through the center of their villages, and left only a few security detachments on the outskirts facing the enemy. Permanent structures destroyed by artillery fire or aerial bombs were utilized as defense points. The ruins hid weapons and served to strengthen the underlying bunkers. Even the heaviest shelling would not drive the Russians from such positions; they had to be dislodged with hand grenades or flame throwers. The Russians upon retreating frequently burned or blasted buildings suitable for housing command posts or other important military installations. Quite often, however, they left castles, former countryseats, and other spacious dwellings intact, after they had mined the walls in a completely conspicuous manner with delayed-action bombs, which were often set to explode several weeks later. These were meant to blow up entire German headquarters at one time. The possible presence of time-bombs in cities, railroad stations, bridges, and other important structures always had to be taken into account.

When the Russians were on the offensive, they tried to encircle fortified towns in order to bring about their fall through concentric advances. Only during major offensives would the advance forces bypass inhabited places in order to gain ground rapidly, leaving the job of mopping up to the reserves following behind. If the Russians were encircled, they defended themselves very stubbornly, capitulating only in rare cases. The large-scale battles of encirclement of 1941, when hundreds of thousands surrendered in hopeless situations after previous attempts to break out, were exceptions.

Fighting for the possession of villages played a still greater role in the winter. The villages blocked the few roads which had been cleared of snow, and offered warm quarters. Cleared roads and warm quarters are the two basic prerequisites for winter warfare. Therefore, inhabited localities retained their outstanding tactical importance even though they could easily be bypassed by ski troops even in deep snow. Experience had shown that ski and sleigh forces might seriously harass the enemy, but they would never be able to bring about major decisions.

The tactics of winter warfare therefore centered around contests for the possession of roads and inhabited places. In Russia, villages and roads were infinitely more important than they were on the rest of the Continent. In other German theaters of war any one particular road never became a crucial factor, since the well-developed road net always offered a choice of alternate routes. In the East, the possession of a single road often was a life or death matter for an entire army. To be sure, inhabited places were also tactically important in France and along the Mediterranean, and offered welcome shelter. Properly clothed, however, the troops were able to remain in the open for a long time without freezing, or even endangering their health—an impossibility in the East. The extreme tactical importance of inhabited places during the 6 months of winter explains the fact that the Russians frequently would much rather destroy them than surrender them to the enemy.

II. Forest Fighting

The Russians favored forests for their approach marches and as assembly areas for an attack. They came and disappeared invisibly and noiselessly through the woods. Narrow strips of woodland leading to the outskirts of villages were used as concealed approaches by reconnaissance patrols. The woods also indicated the logical course to be followed for the forward assembly prior to an attack as well as for infiltrating into German positions. Outskirts of woods were a preferred jump-off position for the Soviet mass attacks. Wave upon wave would surge out of the forests. Undaunted by the losses that the German defensive fire inflicted on their ranks, the Russians continued the attack. Even small clearings were used for artillery firing positions. If necessary, the Russians would create such clearings by rapidly felling trees. Quickly and cleverly they constructed positions for heavy weapons and observation posts in trees, and so were able to lend effective support to their advancing infantry. Bringing up even medium artillery and tanks through almost impenetrable forests presented no problem to the Russians.

In June 1944, Russian tanks reached a trackless forest east of Lwow through a narrow gap in the front. The whole tank corps soon followed, although it was attacked from both flanks and heavily bombarded by artillery and rocket projectors. The Russians used their heavy KV1 and KV2 tanks as battering rams to crush the medium growth of timber. The attached engineers overcame some of the attendant difficulties by laying corduroy roads across the swamps, and the infantry and artillery were soon able to follow the tanks. Shortly before the Russian operation, the commanders of the German panzer divisions had come to the conclusion that this forest was impenetrable even for Russian tanks. The Russian advance over this hastily improvised road, constructed with the aid of the most primitive facilities, was, for a time, accompanied by the strains of band music!

In an attack across open terrain with only occasional patches of forest, the Russians endeavored to reach those patches in the shortest possible time. The Germans found that forests had the same magnetic attraction for the Russians as inhabited places. Whenever the Russians planned a river crossing, one could safely assume that it would take place where woods or inhabited localities reached down to the banks of the river.

When the Russians in the course of their great counteroffensive successfully effected a break-through 60 miles deep, west of Byelgorod on

5 August 1943, they seriously threatened the flanks and rear of XI Infantry Corps on the upper course of the Donets River. The Russians recognized the critical situation of the German corps and sought, by a thrust across the river, to cut its only route of retreat. In spite of heavy losses, the Russians managed to gain a foothold in a forest on the west bank of the Donets. On that occasion, a ruse paid the Russians handsome dividends. When the German local reserves launched a counterthrust, they suddenly faced a Russian battalion dressed in German uniforms, and immediately ceased firing. By the time the German troops became aware of their mistake, it was too late. The Russians took advantage of the resultant disorder, fell upon the deceived attackers, and took a large number of them prisoner. The Russians then entrenched themselves in a larger patch of woodland in order to continue their thrust from there. A counterattack, begun the next day and supported by the massed artillery of the German corps and a rocket projector regiment, succeeded in compressing the Russians into a small area, but not in driving them back across the river, although the concentrated drum fire killed three-quarters of their forces.

The innate aptitudes of the Russian soldier asserted themselves to an even greater degree in defensive actions fought in forests. The Russian command was very adept at choosing and fortifying forest positions in such a way that they became impregnable. On the edge of woods toward the enemy, the Russians left only outposts for guarding and screening the main line of resistance, which was withdrawn deep into the forest itself. That security line also formed the springboard and the support for reconnaissance, scouting, and other operations. The main line of resistance frequently ran parallel to the opposite edge of the woods and a few hundred yards inside the woods. Very extensive woods often concealed groups of bunkers in the central part. These bunkers, constituting an intermediate position, were to delay the advance of the enemy, deceive him as to the location of the main position, and serve as support for the outposts. The Russians also protected the exposed flanks of a forest position by groups of bunkers. Important approach routes were blocked by individual machine-gun or antitank-gun bunkers, echeloned in depth. The immediate vicinity of the bunkers was protected by mined entanglements of branches and abatis, as well as by snipers in trees. Furthermore, the Russians used to mine all bypasses and clearings in numerous places. These measures greatly delayed German progress through a forest, because the bunkers could be taken only after costly fighting, and because engineers had to be called upon for time-consuming mine-clearing operations. Important forward strong points in forests had facilities for all-around defense. A forester's house or a hamlet would often form the central point of the fortified position. A defense trench surrounded by obstacles and mine fields completely encircled the position. The few sally ports were guarded by sentries and movable barriers. A ring of bunkers, connected with each other and with the fortress, enclosed the central point. The intermediate position was blocked by barbed wire, entanglements of branches, and mines. The previously described individual bunkers were placed along the approach routes.

An extensive system of bunker groups formed the battle position and made possible an unbroken defense of the front. In the battle position all the previously described defense expedients were found in even greater numbers. Entanglements of branches interwoven with barbed wire, and mined abatis to a depth of several hundred yards were no rarity. These obstacles prevented sudden thrusts along the roads. Wherever there might have been a possibility of bypassing these obstructions, one could be sure that mines or tank traps had been installed. In such cases the German troops would often end up in a swamp or an ambush. All bunkers and defense installations were so well camouflaged that they could never be discovered by aerial reconnaissance, and by ground reconnaissance only at very close range. Because of the system of advanced strong points and security positions, it was in many cases impossible for German scouting parties even to get close to the main defensive position. While reconnaissance in force by at least a reinforced battalion might succeed in breaking through the outer protective screen, it frequently would bog down at the supporting position of the Russian outposts. However, if the Russian gave way without offering much resistance, the utmost caution was indicated since a further advance was sure to end in a prepared ambush. In such cases, entire German companies repeatedly were wiped out, the prisoners being massacred.

After the capture of Sukhinichi in February 1942, the German 208th Infantry Division continued its northward thrust. The continuous threat from Red forces in a woodland, as well as raids on the supply road of the division, made it imperative that the woods be cleared of the enemy. According to reports from scouting parties, a small village occupied by a substantial number of Russians was located in the center of the woods. Since it was to be presumed that the raids were launched from that point, the division ordered the capture of the village. Half a battalion was assigned to the task. The force was partially equipped with snowshoes and reinforced by infantry heavy weapons which were taken along on sleds. The Germans advanced along the road leading from Sukhinichi to the village and, having arrived at a clearing in the woods—in the center of which lay the village—without being molested, surrounded the village. The attack met bitter resistance and failed. After having sustained heavy losses, the half-battalion retreated along its approach route. Meanwhile, Russian troops had taken up positions along the road and fell upon the retreating Germans. Only remnants of the half-battalion reached their point of origin.

Strongly garrisoned Russian forest positions were difficult to attack and always cost many casualties. They were invulnerable to attacks by the Luftwaffe, German artillery, or armor. At best, tanks and self-propelled assault guns could be employed individually or in small groups, in which case they were very useful. Very rarely could strong positions be taken if they lay deep in an extensive forest or near its far edge. Frequently whole German divisions were pinned down before such positions until they could be relieved from their plight by an envelopment by other forces. For that reason, forward thrusts by strong forces were not led through woods but around them, wherever such a maneuver was possible.

In the latter half of August 1941, the German 6th Panzer Division was to begin a thrust toward Leningrad from the Porechye bridgehead on the Luga River. (Map 4) The bridgehead was completely surrounded by woods and the sector to be attacked lay in a medium-growth, partly marshy woods with thick underbrush. The sector was occupied by the 2d and 3d Russian Proletarian Divisions. The most advanced Russian position was located about 300 to 400 yards ahead of the German front. The Russian trenches were narrow and deep, and had no parapets. The excavated earth had been scattered in the surrounding rank marsh grass, and the trenches were so well camouflaged with branches that neither reconnaissance patrols nor aerial photography had been able to spot them during the preceding 4 weeks of fighting. The wire entanglements were no higher than the dense growth of grass hiding them. Single roads from the southwestern and northeastern ends of the bridgehead cut through the woods to a village beyond. The two roads were blocked by heavily wired abatis and mine fields. On the far edge of the woods a second position was located atop a sand dune, a third ran through the village, and a fourth lay behind the village. The second position was particularly well constructed. It consisted of a deep antitank ditch, in the front wall of which the Russian riflemen had entrenched themselves, and bunkers for heavy weapons had been installed.

The German attack was to be launched along the two above-mentioned roads. A reinforced armored infantry regiment, supported by strong artillery elements and a rocket projector (Werfer) battalion, advanced along each of these roads. Individual tanks were to support the engineers in the removal of the road blocks. In spite of very heavy fire concentrations on the projected points of penetration, the Russians could not be budged from the narrow, invisible zigzag trenches. To be sure, the German tanks were able to reach the barriers, but the dismounted engineers were unable to remove the blocks in the defensive fire, which continued unabated. The infantry following up sought fruitlessly to find other weak spots in order to effect a break-through. Repulsed everywhere by the murderous defensive fire of an invisible enemy, it finally stopped, knee-deep in swampland, before the wire entanglements in front of the still unknown Russian position. Not until the following night did one German company succeed in crawling forward, man by man, through the deep-cut bed of a brook which was overgrown with grass and bushes, and in infiltrating through the entanglement. That particular point had not been attacked theretofore. Strong German reserves were immediately brought up. They widened the point of penetration, and cleared the trenches and strong points of the westerly sector after hours of hand-grenade fighting. The Soviets continued to maintain their position in the eastern sector. German forces could be directed against the rear of the enemy only after a thrust into the depth of the western sector had reached and rolled up the second position. After bitter hand-to-hand fighting the German forces were finally able to scatter the Russians also at that point, and to clear the road. Only then, after a 2-day battle which exacted a heavy toll of losses from both sides, could this invisible defensive system in the woods be surmounted.

III. Fighting Beside Rivers, Swamps, and Lakes

During the course of the war, the ability of the Russians to cross even the largest rivers was always a source of amazement to the Germans. When the German armies reached the Dnepr in the summer of 1941, the problems of surmounting the obstacle presented serious difficulties to the German command, since they had no previous conception of the size and the nature of the river. How quickly and easily this problem could be solved was demonstrated a few days later by the Russians. During the course of one night a cavalry corps crossed the river—using field expedients for ferrying men and equipment to the opposite bank, and swimming thousands of horses across—and penetrated deep into the lines of the surprised Germans.

Two years later the German armies retiring westward headed for this same sector of the Dnepr and had great difficulties in reaching and crossing it ahead of the Russians. By calling on all the forces and means at their disposal, the Germans had managed to occupy seven existing bridges in a sector 300 miles long, but were able to establish only one float bridge and one improvised ferry because of the scarcity of ferrying equipment that prevailed by that time. On the other hand even before the German troops arrived, the Russians, following in close pursuit, succeeded in dropping several thousand paratroops over a 200-mile-long part of the sector, in establishing small bridgeheads at several places, and soon thereafter in building 57 bridges, 9 foot bridges, and other facilities for crossing the river. Thus, the Germans had a crossing over the Dnepr every 40 miles, the Russians one every 4 miles. At one ferrying point 25 miles downstream from Kremenchug, the Russians established a small bridgehead and proceeded to ferry tanks across the river on rafts, by day and night. Their operations continued even when they were shelled by German artillery, and some of the tank-laden rafts went to the bottom of the river.

The Russians also made extensive use of raft bridges built by their engineers. The German engineers first learned about these from the Russians and had to experiment in order to determine their load capacity. Raft bridges could be used only for crossing waters having a slow current. The raft bridge is built of tree trunks placed side by side and fastened to each other. Depending upon requirements, a second and even a third layer of logs is added, each layer being laid at right angles to the layer below. Planking, laid across the uppermost layer, serves as a roadway. The load capacity of the raft bridge can be adapted to meet existing requirements by varying the number of layers of logs. The Russians built bridges of this type ranging from a 5-ton bridge at Rogachev in 1941 to a railroad bridge of over 100-ton load capacity near Kiev in 1943. Just 4 days after taking Kiev, the Russians had established rail communications into the city over this heavy raft bridge across the Dnepr.

Swamps and lakes also presented no real obstacles to the Russians as was demonstrated during the battles on the Volkhov and the engagements on the Luga and Desna, as well as in many other swamp sectors. When, in February 1944, a yawning gap opened on the Pripyat River at the boundary between Army Group North Ukraine and Army Group Center, the Russians crossed this extensive marshy region during the muddy period with 14 divisions, and pushed toward Kowel. Several of these divisions turned south through Rowne to attack Lwow. Stopped near Dubno on the Ikwa River by the German Fourth Panzer Army, they vainly tried to take the few strongly manned crossings over the extensive swamps on both sides of the river. Nevertheless, one morning a Russian battalion appeared in the rear of Dubno. It was surrounded by German armor and captured. Interrogation of the prisoners revealed that during the night the Soviet riflemen had crawled on their bellies across the slightly frozen marsh, which was up to 600 yards wide and could not be crossed on foot, and arrived exhausted and covered with muck.

In September 1943, the German XI Infantry Corps stood on the Dnepr astride Kremenchug, and had to protect its right wing against the enemy who had broken through in the adjacent sector. A shallow lake, 2½ miles long and from 300 to 500 yards wide, facilitated the flank protection. The western bank was guarded by weak German forces. One night these forces were suddenly attacked and driven back by from 600 to 800 Russians. Under cover of darkness the Russians had waded across a shallow spot of the marshy lake without a stitch of clothing, and—equipped only with small arms and ammunition—had surprised the German security. Only quickly brought up mobile reserves were able to encircle the Russians and take them prisoner.

IV. Fighting in Darkness and Inclement Weather

The Russians used darkness and fog primarily for troop movements, preparations for attack, construction of field fortifications, and supply operations. Reconnaissance in force, and raids likewise, were usually carried out under cover of darkness or hazy weather. In these instances, the Russians proceeded with patience, cunning, and perseverance. Not infrequent were Russian night attacks in strength of up to a regiment.

In the Arctic, the Russians employed commando teams specially trained in Byelomorsk for night raids into the German rear area. Appropriate to the nature of their mission, these troops were equipped only with absolutely essential items. Minutely detailed orders took care of every phase of the undertaking, and were carried out methodically and to the letter. Here, too, the Russian proved himself a fearless fighter. As in other regions, attacks on strong points in the Arctic were broken off only when the Russian casualties amounted to many times the strength of the strong point complements.

A typical example of a planned night attack was that of a Siberian division against the German 112th Infantry Division southeast of Uzlovaya toward the end of November 1941. (Map 5) After it had finished unloading, the Siberian division advanced on the 112th Infantry Division early in the night. Having gained a substantial amount of territory to the north beyond Bogoroditsk in minor engagements during the previous day, the 112th Division had bivouacked for the night. About 20 tanks led the Siberian attack. The mere appearance by night of tanks in front of the lines of the 112th Division produced a severe shock. No means of defense were on hand at that time. Any defenses would have had only a local effect at night. When the attacking Siberians appeared behind the tanks, complete panic ensued. The elements of the 112th Division hit by the attack fell back several miles, close up to the northern outskirts of Bogoroditsk. Special steps had to be taken to restore control of the situation. The territorial gains of the Siberian division remained limited to a few miles; a large-scale exploitation of their success did not follow. The Russians probably had reached their objective and had not planned any further advance.

Night attacks on a major scale, however, remained the exception to the rule. The Russians undertook such attacks only when they had orders from a higher command, or when they failed to take an important objective in a day attack in spite of a mass commitment of men and equipment. Night attacks were generally acts of desperation, where everything was staked on one card.

During the polar winter only small-scale warfare was possible which, however, the Russians as well as the Germans waged zealously. Security and reconnaissance activity had to be greatly increased. Surprise attacks were always to be expected because it remained dark throughout the day. The generally known principles of defense against night operations held true under those conditions. Almost the only new feature was the fact that fighting was not limited to mere local actions, for raiding parties consisting of smaller units did not hesitate to thrust far into the depths of the enemy front (the Russians to the Turku- and Helsinki-Petsamo highway (Eismeerstrasse), the Germans to the Murmansk railroad). Both sides frequently employed specially trained troops in these operations. The Russians used troops trained in Byelomorsk; the German used special Finnish units.

Attempts to break out of pockets, such as took place in 1941 at Maloryta, Vyazma, Bryansk, and other encirclements, were another type of night surprise attack. Here, the Russians made no long preparations, but hurled themselves into the open without any definite plan. In tightly packed hordes they struck at whatever points they believed there was a chance to break through. At Maloryta the Russians, after a successful breakout of part of their forces, spent 36 hours encircling and storming a village still held by the Germans. This village was to the rear of the Russians, and by their pointless operation they lost the advantage of their nighttime breakout. Possession of the village would have been completely unimportant for the success of the breakout. Their orders, however, probably read that way.

Russian large-scale offensives always started by daylight. Usually the early morning hours were chosen. H Hour remained the same even if dense fog obscured everything all day long as, for instance, at the beginning of the second battle of East Prussia on 14 January 1945.

It was, however, highly advantageous for the German to attack heavily fortified Russian strong points, and antitank gun and tank fronts by night. Attempts to take them by day would have cost large numbers of lives. Night attacks were almost always successful when carried out by troops specially trained for this type of combat, and usually cost only a few casualties.

In the spring of 1943 during the war of position north of Tomarovka, an assault detachment in company strength of the German

167th Infantry Division succeeded in infiltrating the Russian front by night, raiding a strongly garrisoned village from the rear and driving the enemy out. During the previous day, two battalions had been unable to take the village despite strong artillery support.

In January 1943 a company of the 6th Panzer Division supported by six self-propelled assault guns was similarly able to take an important fortified village north of Tatsinskaya by a night assault from the rear. Previously, a panzer grenadier regiment and a 60-tank panzer battalion of the neighboring division had vainly endeavored all day to take this village.

Verkhniy-Kumskiy, the bitterly contested key point of Russian defense against the German relief thrust on Stalingrad, also was taken in a night attack by a panzer grenadier battalion with but minor losses.

Even the Russians began no far-reaching operations in really bad weather, but such weather suited them very well for local operations. During fog and blizzards the Russians always developed lively reconnaissance activity and raided advanced security posts. Not infrequently the Russians would attack in battalion or regimental strength during driving thundershowers to effect reconnaissance in force, to improve their position, or to gain a favorable jump-off position for a major offensive. In winter they exploited the cold eastern storms of the steppes for such assaults, especially in the southern sector. On those occasions the Russians often succeeded in entering the German trenches without firing a shot, and in taking many prisoners. Indeed, the Russians knew very well that the easterly gales drove such clouds of powdered snow ahead of them that the German soldiers were unable to observe and take aim against the wind. They were, therefore, practically defenseless. Only by a ruse were the German divisions fighting there able to regain mastery of the situation. Those front sectors particularly threatened during the easterly gales were simply evacuated and the forces quartered in the villages situated along the sides of the gaps. When the Russians rushed forward into, or over the empty trenches, the German forces wheeled against the rear of the Russians and attacked them from the east; the Russians were then just as defenseless as the German had previously been and were often captured *en masse.* As a result, the Russians later ceased attacking during easterly gales.

Chapter 13

Camouflage, Deception, and Propaganda

Camouflage, deception, and propaganda were expedients much used by the Russians. These, too, reflected in every aspect the oriental character of the people. The Russians carried out measures conforming to their natural talents, such as camouflage and deception, with great skill and effectiveness. Their front propaganda, however, was crude and naïve for the most part. Because it did not correspond to the psychology and mentality of the German soldier in any way, it was ineffective. Although pursued zealously, and with a great variety of media, it attained no appreciable success for precisely those reasons.

I. Camouflage

The Russians were excellent at camouflage. With their primitive instinct they understood perfectly how to blend into their surroundings and were trained to vanish into the ground upon the slightest provocation. As illustrated in the preceding pages, they skillfully used darkness, vegetation, and poor weather conditions for concealing their intentions. Their movements at night and their advances through wooded terrain were carried out with exemplary quietness. Now and then they would communicate with each other by means of cleverly imitated animal cries.

Noteworthy, too, was their camouflage of river crossings by the construction of underwater bridges. For this purpose they used a submersible underwater bridging gear, which could be submerged or raised by flooding or pumping out the compartments. The deck of the bridge was usually about 1 foot below water level, and was thus shielded from aerial observation.

Artificial camouflage was another device used by the Russians. Even at the beginning of the war the Germans came across Russian troops wearing camouflage suits dyed green. Lying prone on the grass, these soldiers could be spotted only at a very short distance, and frequently were passed by without having been noticed at all. Reconnaissance patrols frequently wore "leaf" suits of green cloth patches, which provided excellent camouflage in the woods. Russians wearing face masks were no rarity.

The Russians enforced strict camouflage discipline. Any man who left his shelter during the day was punished severely, if it was forbidden for reasons of camouflage. In this way the Russians were able to conceal the presence of large units even in winter, as the following example illustrates.

In January 1944 the Russian First Tank Army attempted to take the important railroad hub of Zhmerinka in a surprise attack. The attack was repulsed, and the army encircled to the southeast of Vinnitsa. It broke out of the encirclement during the very first night, and disappeared. The bulk of this Russian army escaped completely unnoticed through the gaps left by the insufficient German forces engaged in the encirclement. In spite of deep snow and clear weather, it could not even be determined in which direction the Red forces had escaped. From the situation, it was to be assumed that they had hidden in the immediate vicinity in a group of numerous, rather large villages with extensive, adjoining orchards. Since German armored units had previously driven through those villages, the tank tracks gave no reliable evidence that the Russians were hiding there. For 2 days and nights the Luftwaffe scouted for the whereabouts of the First Tank Army, and in this connection took excellent aerial photographs of the entire area in which the villages were located. But neither aerial observation nor the study of aerial photographs provided any clue. Not until the third day, when a strong German tank force pushed into the group of villages, was the hiding place of the entire army established in that very area. All tanks and other vehicles had been excellently camouflaged in barns, under sheds, straw piles, haystacks, piles of branches, etc.; and all movements during the day had been forbidden, so that nothing gave away their presence.

II. Deception

Prior to offensives the Russian made extensive use of deception. In order to mislead the enemy as to the time and place of impending large-scale offensives, the Russians feigned concentrations in other sectors by preparing a great number of fire positions for artillery, mortars, and rocket projectors. They strengthened this impression by moving smaller bodies of troops into those sectors by day and night, as well as by setting up dummy artillery pieces, tanks, and aircraft, and making appropriate tracks leading up to them. The Russians also were known to place an entire tank army behind an unimportant sector, so as to create the false impression of an impending attack from that point. By running their motors at night they sought to create the impression that tank and motorized columns were on the move. Artillery trial fire and the use of roving guns likewise were among the most commonly used Russian deceptive practices.

For purposes of deception on a more limited scale, the Russians frequently used German uniforms for whole units as well as for individuals. That method of deception was almost always successful.

In the summer of 1943 a German-speaking Russian in the uniform of a German officer succeeded in driving a German truck right up to headquarters of the Rowne military government detachment (Ortskommandantur), and in obtaining an audience with the commandant, a general. He gagged the commandant, wrapped him up in a big rug, carried him out of the truck which he had left idling outside, and delivered him to the partisans. Only by the words, "Thanks, comrade"—words that an officer in the German Army simply would not use in addressing a private—did he arouse the suspicion of the kidnapped officer's orderly, who had innocently helped him load the heavy carpet into the truck. Although an immediate report was made by the orderly, it did not lead to the apprehension of the kidnapper.

In Lwow, in the spring of 1944, apparently the same Russian, dressed as a German officer, succeeded by similar trickery in killing the deputy governor of Galicia as well as a lieutenant colonel, and later a sentinel who wanted to inspect his truck. Each time he succeeded in escaping.

Similar surprise raids and deceptions of combat troops through the misuse of German uniforms occurred at all sectors of the front in ever increasing numbers.

III. Propaganda

The propaganda of the Russians exploited military and political problems and used all the technical means by which modern propaganda is disseminated—radio, press, leaflets, photographs, planes flying by night towing illuminated streamers or equipped with loud speakers and photographs, loudspeakers set up on the ground, leaflet shells, rumors spread by agents, Russian PW's, and Germans pretending to have escaped from Russian captivity. Since there were also Finnish troops on the Arctic Front, to whom different things were of importance, Russian propaganda at that front sometimes met with difficulties, and was not properly coordinated. One also had to differentiate between propaganda aimed at higher military commanders, the troops, and the German people. The various groups to be propagandized were dealt with and approached from entirely different angles. From a literary and artistic standpoint, much of the Russian propaganda was of high caliber.

The Russian intelligence service covered events in the German Army with amazing speed and accuracy. Photographs, for example, taken by one of the German propaganda companies for various reasons, appeared almost simultaneously in the pertinent Russian army newspapers. The propaganda at the front, however, was crude and clumsy. For that reason, it made but little impression. Political and military satire was used a great deal.

The Germans obtained information on Russian propaganda addressed to their own troops from captured Russian army newspapers. It, too, employed words and pictures. Nationalism and ideological fanacticism were exploited with equal intensity. The Russians seemed to criticize quite frankly some of the events at the front. Conditions among German troops in opposing positions were treated satirically for the most part. In the propaganda directed at their own front lines, the Russians pounded into their soldiers' heads the story that the Germans shot every PW on the spot. This propaganda-induced fear of being taken prisoner was to make the Russians soldiers stand their ground to the very end. The story was believed, and that particular line of propaganda accomplished its purpose. There were relatively few Russian deserters. On the other hand, older Russian soldiers who had worked in Germany as prisoners during World War I were immune to such atrocity stories and deserted very frequently.

A large part of the Russian propaganda effort was devoted to studying and counteracting German propaganda activities. Reading and passing on German propaganda leaflets was forbidden under penalty of death.

By erecting signs, by loudspeaker messages, or by dropping propaganda leaflets from aircraft, the Russians made extensive use of front-line propaganda urging the Germans to surrender or desert. Often the commander was addressed personally, sometimes by captured German officers of all ranks, who allegedly belonged to the "Free Germany" organization. Many propaganda leaflets were dropped that represented pictorially the Russian superiority in men, weapons, and matérial, as well as in armament potential. Also alleged German atrocities and acts of destruction were shown. It was, however, not quite clear just what propaganda purpose the exhibition of nude women standing on the breastworks of Russian trenches was supposed to accomplish. They supposedly were German girls who had fallen into Russian hands.

The German command had little reason to be concerned over Soviet propaganda. Neither the spoken nor the printed Russian propaganda inspired and credulity, for it contained too many obvious lies. Besides, the German soldier had seen the dubious blessings of Bolshevism at close range, and had discussed this dictatorial system with its opponents among the Russian people. There were a great many anti-Communists among the Russian intellectuals. Those people voluntarily joined in the German retreat in 1943, because they wanted to have nothing more to do with the Russian system.

Even before the Eastern Campaign, strong anti-German propaganda was disseminated. In the schools of many Russian cities and villages German language texts were found which contained the most coarse invectives aimed at Germany. That was the manner in which communism spiritually prepared for war against Germany. In a fairly large village south of Leningrad, half-grown, German-speaking boys naïvely admitted that they had been selected as *Komsomoltsy* (the Soviet counterpart of the German Hitler Youth) for Magdeburg.

Occasionally the Russians also doubled back German PW's with false reports. The Germans frequently picked up particularly well-trained Russian deserters who were supposed to supply them with false information. This type of propaganda was somewhat less than a success. With the exception of individual non-German soldiers in German uniform, instances of desertion from the German rank remained limited to the acts of a few desperadoes.

During the first years of the war the Russians apparently had sought to impress the German troops and lower their morale by committing numerous atrocities against them. The great number of such crimes, committed on all sectors of the front especially in 1941–42, but also during later German counteroffensives, tends to support that presumption.

On 25 June 1941, two batteries of the German 267th Infantry Division near Melniki (Army Group Center) were overrun in the course of a Russian night break-through and bayonetted to the last man. Individual dead bore up to 17 bayonet wounds, among them even holes through the eyes.

On 26 August 1941, while combing a woods for enemy forces, a battalion of the German 465th Infantry Regiment was attacked from all sides by Russian tree snipers, and lost 75 dead and 25 missing. In a follow-up thrust, all of the missing men were found shot through the neck.

In January 1942, an SS division attacked the area north of Szyczewka (Army Group Center). On that occasion, a battalion fighting in a dense forest area suffered a reverse and lost 26 men. German troops who later penetrated to that point found all the missing SS men massacred. In April 1942, an elderly Russian civilian, a carpenter, appeared at a German division headquarters southwest of Rzhev, and reported that he had encountered a group of about 40 German PW's with a Russian escort in his village a few miles behind the Russian front. The prisoners, he continued, had soon afterward been halted at the northern outskirts of the village, where they had dug deep pits. According to eyewitness reports, the prisoners had subsequently been shot, and buried in those pits. A few days later, the village was captured in a German thrust. The incident was investigated, and found to be true.

During the battle of Zhizdra, in early March 1943, a battalion of the German 590th Grenadier Regiment was assigned the mission of mopping up a sector overgrown with brush. The attack failed.

When, on 19 March 1943, the sector again passed into German hands after a counterattack by the corps, 40 corpses of soldiers from the battalion were found with their eyes gouged out, or their ears, noses, and genitals cut off. Corpses found in another sector of the battlefield bore signs of similar mutilations.

On 5 July 1943, during the German pincers attack on Kursk (Operation ZITADELLE), a battalion on the southern flank of the 320th Infantry Division lunged forward without being supported by other units. It ran head-on into the counterattack of an enemy division, and was repulsed. About 150 men were taken prisoner. Shortly thereafter the Germans monitored a telephone conversation between Russian lower and higher headquarters (probably regiment and division), which went about as follows:

> Regimental commander: "I have 150 Fritzes (derogatory terms for German soldiers) here. What shall I do with them?"
>
> Division commander: "Keep a few for interrogation, and have the others liquidated."

In the evening of the same day, the presumed regimental commander reported the order executed, stating that the majority of the Fritzes had been killed immediately, and the remainder after they had been interrogated.

The Russians sought to intimidate their own civilian population by similar atrocities. In March 1943, after the recapture of Zolochev, a small city 20 miles north of Kharkov, the inhabitants told the German military police that the Russians, before their retreat, had herded and whipped a rather large number of local boys between the ages of 14 and 17 years naked through the streets in intense cold. Afterward, they were said to have disappeared into the firehouse, where the NKVD had its headquarters, never to be seen again. During a subsequent search, all of the missing boys were found in a deep cellar of the firehouse, shot through the neck and covered with horse manure. The bodies were identified and claimed by relatives. All had severly frostbitten limbs. The reason for this particular atrocity was assumed to have been the alleged aid rendered the German occupation forces.

One medium of propaganda employed frequently by the Russians during the last years of the war were German PW's who would be doubled back to the German lines, usually to sectors held by their own regiments, with the mission of inducing their comrades to desert by telling them how well they would be treated as Russian prisoners. That type of propaganda failed, as did similar attempts on the part of the Germans with Russian PW's who had volunteered for this assignment.

Few Russian soldiers, on the other hand, believed German propaganda. Whenever it fell on fertile soil, its effects were promptly neutralized by counterpropaganda and coercion. Except during the great encirclement operations, there were only isolated instances of wholesale desertion of Russian units. If it did occur, or if the number of individual deserters increased, the Soviet commissars immediately took drastic countermeasures.

During the protracted period of position warfare along the upper course of the Donets in the spring of 1943, a front-line unit of the German XI Infantry Corps south of Byelgorod was able to take a large number of prisoners. The prisoners were taken in midday raids, since it had been ascertained from deserters that the Russians in this terrain sector—which could be readily observed from the western bank of the river—were allowed to move only at night, and therefore slept during the day. The prisoners admitted that many of their comrades were dissatisfied and would like to desert; however, they were afraid of being fired upon by the Germans and would have difficulties crossing the deep river to the German lines. Contact with the company of malcontents was soon established, and the necessary arrangements made. Unobstrusive light signals on the chosen night informed the Russian company that the necessary ferrying equipment was ready, and that German weapons stood ready to cover their crossing. All necessary precautions had been taken in case of a Russian ruse. Just the same, the company really dribbled down to the banks of the river, and in several trips was ferried across the Donets in rubber boats; the company commander, an Uzbek first lieutenant, being the very first. Part of the company, however, ran into Russian mine fields, suffering considerable losses from exploding mines as well as from the fire of the alerted Russian artillery.

The result of this undertaking and the above-mentioned incidents was that, having become unreliable, the 15th Uzbek Division was immediately withdrawn from the front, disciplined, and committed elsewhere.

PART FOUR

THE RED AIR FORCE

Chapter 14

A Luftwaffe Evaluation

Numerically, the Red Air Force was greatly underestimated by the Germans before the beginning of the Eastern Campaign. But in spite of its numerical strength, which increased considerably during the course of the war, it had no decisive influence on the outcome of the battles in the East.

Russian air force tactics were inflexible and strictly followed a fixed pattern. They were wanting in adaptability. Only in late 1944 and early 1945 could the first beginnings of strategic air warfare be observed. The Russian Long Range Force (*Fernkampffliegerkorps*), which came under the surveillance of German radio intelligence as early as 1941, was employed primarily in transport operations. Although the Red Air Force was an independent service of the Russian armed forces, it was employed almost exclusively on the battlefield, in joint operations with the army.

The Germans detected impending Russian attacks by—among other clues—the early assembly and concentration of combat aviation on airfields near the front. In this connection, the Russians proved very adept at building auxiliary airfields. Ruthlessly exploiting labor forces drawn from the civilian population, and using the most primitive equipment, they would have the airfields completed and ready for take-offs within an amazingly short time. Neither winter nor the muddy periods interfered with their work. The Russian Air Force made liberal use of dummy airfields and aircraft, as well as of numerous methods of camouflage.

In combat, the direction and commitment of aviation was assumed by command posts near the front, one of the most ably handled phases in this respect being the control of fighter aircraft from the ground. On the other hand, cooperation between fighter and ground-attack aircraft or bomber formations left much to be desired. Fighter escorts seldom accompanied them on their missions; if they did, they scattered upon first contact with the enemy.

The Russians proved to be excellent bad-weather pilots. Although not equipped for instrument flying, fighters and ground-attack aircraft hedge-hopped over battlefields in the most inclement weather. They liked to take advantage of low ceilings and blizzards in order to surprise the enemy.

Russian night fighters as a rule confined themselves to attacks on those targets on which they had been briefed, and were equipped with only the most basic navigation aids. The Germans were amazed to discover that Russian night fighters almost always flew with their position lights burning.

Air force formations concentrated for major operations always revealed a rapid decline of fighting potential once they had joined action. The number of planes capable of flying combat missions decreased rapidly, and a rather long time was required to restore them to flying condition. The Russians made extended use of artificial smoke for camouflaging and protecting industrial plants, railroad junctions, and bridges against strategic air attacks.

The rapid repair of bomb damage was noteworthy, especially in the case of railroad installations. Again, labor forces for this purpose were ruthlessly commandeered from the civilian population.

Air supply operations assumed substantial proportions during the course of the war. Planes either landed the supplies, or dropped them by parachute. Agents and armed saboteurs dropped behind the German lines likewise played a special role.

At the beginning of the war, Russian ground troops were extremely vulnerable to air attack. Very soon, however, a change took place. Russian troops became tough and invulnerable to attacks by German dive-bombers and ground-attack aircraft.

Training in defense against low-level attacks was well handled. Every weapon was unhesitatingly turned against the attacking aircraft, thereby constituting a formidable defense. Whenever the weather permitted, the Russian troops avoided billets and concealed themselves masterfully in the terrain. If, in exceptional cases, they sought shelter in inhabited places, they had strict orders not to show themselves outdoors in the daytime.

In keeping with Soviet ideology, the Russians employed an increasing number of female pilots and other female air crew members as the war went on. Women not only flew transport missions, but manned combat planes as well.

In conclusion, it may be said that the Red Air Force, although conceived and built up on a large scale, was very primitively trained. Its will to fight, its aggressive spirit, and its mastery of technical aspects left much to be desired. Although constantly superior in numbers to the Luftwaffe, it was always inferior when it came to combat. Usually a small number of German fighters sufficed to clear the skies of Russian planes.

Chapter 15

A Ground Force Evaluation

I. Tactical Employment

With regard to matériel and training, the Soviet Air Force in World War II was always much inferior to the Luftwaffe despite the fact that the number of its planes increased steadily and had outnumbered those of the Luftwaffe as early as 1942. Not even the introduction of new Russian types of planes was able to effect any decisive change in the disproportionate performances of the belligerents. For that reason, the Russian Air Force often was no factor at all in ground warfare. Sometimes it played a secondary role, but never one as decisive as that of the air forces of the Western Allies.

At the beginning of the great German offensive in the East, the ground troops saw only flights of three or four reconnaissance planes, individual bomber squadrons, and only a few fighters (Soviet I–16 single-seat pursuit planes, dubbed "Rata" in the Spanish Civil War). They quickly became victims of German fighters. Seldom did one of the reconnaissance planes return from a mission. No sooner had they been sighted than a long trail of smoke told of their annihilation. Their crews followed them down in parachutes, they being among the first Russian prisoners who with bitter hate, or, in individual cases, with uncontrolled sobbing, awaited their fate: they expected to be shot, as their propaganda had led them to believe. They became all the more confused when the Germans treated them in a friendly manner. The same thing happened to the bombers which flew straight toward their targets, without fighter escorts, in single squadrons of five or six planes, or in from two to three squadrons, one right behind the other. They did not change their course even when the powerful German Flak played havoc with them. Direct hits frequently tore planes into shreds. The rest of the squadron would continue toward its objective until it was shot down by German fighters during the bombing run or the return flight. It happened repeatedly that a whole squadron would be shot down in a few minutes. At that time, an attack by Russian bombers meant nothing more to the German ground troops than an exciting spectacle, which always ended in tragedy for the Russians. Not until the Russians had realized the futility of their efforts did they attempt to ward off their fate by jettisoning their bombs and quickly turning tail upon the approach of German fighters. In that way at least some of their planes succeeded in reaching the home base, only to be finished off the next day.

The Russians were in a fair way to lose their last aircraft by this completely futile commitment in detail. True, the Russian Air Force was able to replace the lost planes, but it never did recover from the shock effect of the German fighters. The superiority of German fighters over Russian planes of any type was evident right up to the end of the war.

Soon, however, the bloody lessons began to bear first fruit. During the battles on the frontier, the Russian planes could be neutralized almost completely, but later, when the German troops were crossing the Dnepr and Dvina Rivers, concentrations of bomber units with fighter escorts did make their presence felt. Skillfully maneuvering, they hit the bridges and crossing sites in surprise attacks; coming in from the flanks or from the rear they harassed the German troops crossing the rivers and were responsible for the first German losses. However, the German crossings of the two rivers were not stopped nor even delayed. At the Luga the Russians employed their new technique by using all aircraft available in the Leningrad area in shuttle raids, in order to destroy individual, isolated bridgeheads of advanced German armored elements. The Germans suffered considerable losses, because their troops were squeezed into a narrow area, and no air support was to be had to counter the Russians because the ground organization of the Luftwaffe had not yet caught up. Nevertheless, the tactical effect was nil, for the Russian Air Force carried on its own private war, as did the artillery, the tanks, and the infantry; of the latter, each division attacked in detail. The result was that in July 1941, for example, the Russian forces, 8 to 10 times superior in strength to the Germans, were not able to take the Porechye bridgehead (70 miles southwest of Leningrad) which had been cut off completely during the first days. The Russians did not succeed in destroying even one of two wooden bridges that were located within 300 yards of each other, after bombing them daily for a period of weeks, and sending 2,000 heavy shells in their direction every day, thus attesting to the meager technical skill of the artillery and the bombers.

The initial lack of any cooperation between the various Russian arms and the deficiency in technical skill, led to continuous failures in the above-mentioned sector as well as on all other fronts. The Russians, gradually recognizing these mistakes, obviously strove for improvement. They developed appreciable technical skill. Cooperation between the combined arms also improved visibly. Nevertheless, this particular aspect remained their weakness to the end of the war. The art of cooperation presupposed a measure of personal initiative, knowledge of tactics, and acute perceptory faculties that the Russian did not seem to possess because those qualities ran counter to his national character and his upbringing. He attempted to compensate for those serious weaknesses by mass commitments of forces and matériel also in the case of his air force. In the air, however, his efforts were not crowned with the same degree of success as on the ground. Eventually, the endeavor on the part of the air force led to the formation of main efforts which became more pronounced from year to year, and which made themselves felt all the more during the Russian large-scale offensives of the last war years when the fighting strength of the Luftwaffe was ebbing visibly because of fuel shortage and grievous losses in other theaters.

The Russian Air Force missed its first great chance in the winter of 1941–42 when the Germans were withdrawing before Moscow. A concentration of most of its air power on the German columns, confined to the few roads that were free from snow, would have had a devastating effect.

Even during the great Russian offensive in the winter of 1942–43 between the Don and the lower course of the Dnepr, as well as during the defense against the German relief push on Stalingrad, the concentrations of Russian aircraft were as yet not strong enough to influence the course of events to any appreciable extent. Air operations properly coordinated as to time and place with those of the ground forces were likewise an exception to the rule.

In the summer of 1943, on the other hand, the concentrations of Russian planes were much greater and their activity much more vigorous. The Russian Air Force played an important role in the battle south of Kursk (the German ZITADELLE attack, July 1943) and in the Russian counteroffensive in August 1943. But its activity soon dwindled and a month later the German armies succeeded in escaping across the few Dnepr bridges to a position behind the river barrier without any interference on the part of the Russian Air Force.

Similar mass sorties were repeated in the battles of Lwow and Vitebsk in 1944, and in the second battle of East Prussia in 1945. The main efforts, however, though clearly recognizable at the start, always dissolved within a short time, and the Russian Air Force would disappear.

II. Combat Techniques

At the start of the war, missions were flown at high altitudes. Fighters (Rata) attacking at low level were first encountered by the Germans on the northern front at the Luga River. They strafed and dropped small bombs on batteries or marching columns, without accomplishing any noteworthy results. During the break through of the Leningrad Line in September 1941, there appeared for the first time small groups of Russian fighters equipped with rocket bombs. Their accuracy and effectiveness was inconsequential.

The IL–2, a very effective and unpleasant ground-attack plane made its debut in February 1943. It could be identified by a cockpit armored with bulletproof glass about 4 inches thick. The pilot was protected by steel plates in the rear and bottom of the cockpit. This type of airplane was invulnerable to rifle and machine gun fire of any caliber. Its armor also withstood 20-mm. Flak projectiles. It is, therefore, understandable that these ground-attack aircraft were used at danger points, and unceasingly harassed ground troops once they had caught them in a low-level attack. In that manner they were able to bring daylight movements of motorized troops to a standstill, and to inflict considerable losses on them with their twin-barreled machine-gun fire and small fragmentation bombs. Flak guns of 37-mm. or heavier caliber were of no use in the defense against their hedgehopping attacks because they flew too fast for allowing proper aim. Thus, it was necessary to use the 20-mm. armor-piercing ammunition, the use of which had previously been forbidden for any other purpose than antitank defense. Once this special-purpose ammunition was used by the 20-mm. Flak, the Russian IL–2's suffered such heavy losses that they rapidly disappeared from the scene.

By the winter of 1944–45, Russian antitank weapons in aircraft had become so highly developed that they represented a seriously growing menace to German tanks and self-propelled assault guns. By January 1945, for example, during the second battle of East Prussia, at least eight self-propelled assault guns were set afire on one day by Russian antitank planes.

That year, too, Russian aircraft began to hunt down German locomotives and individual motor vehicles which they suspected of carrying senior officers. The locomotive hunts led to serious losses on railroad lines located close to the front. The losses ceased only after the trains on those lines had been armed with light Flak. On the Arctic Front, the destruction of locomotives had particularly troublesome consequences. Since the Finnish supply of locomotives was very limited, the Finnish Railroad Ministry wanted to discontinue operations on the Kemijaervi line which was seriously endangered, but urgently needed for German supply operations.

Hunting individual German motor vehicles also led to losses. On Easter Sunday 1944, for example, a German army commander was traveling to the front north of Buczacz, Galicia, when Russian fighter squadrons attacked his car. They made repeated runs on the lone automobile, killed the commander's entourage, blew the car to bits with 80 bombs, and prevented the commander himself from continuing his journey. Only when the higher German commanders used old, nonidentified cars did the attacks cease.

Russian daytime bombing raids even in the later years of the war were carried out mostly by small groups at altitudes of 7,000 feet or higher. Raids by whole wings were an exception to the rule. Carpet bombing by large bomber formations was unknown to the Russians. On the other hand, during the last two war years, they frequently dropped clusters of small fragmentation bombs on live targets. The bombs would fall within a radius of a hundred yards in such a dense pattern that no living object within the effective beaten zone could escape the splinters. The bombs fell into even the narrowest trenches and, because of their great fragmentation, were very dangerous and greatly feared. German planes surprised on the ground often would be set afire or destroyed by the fragments of bombs dropped in such patterns. The effect of these bombs on the morale of the German troops was likewise serious.

The Soviet bombers displayed poor marksmanship. For example, in an attempt to help their sorely pressed infantry on the edge of a forest by bombing the German lines, all the bombs fell on the Russian position and wreaked such havoc that, after the second salvo had been dropped, the German infantry was able to take the position without losses. In most Russian bombing attacks the planes would fly over a troop-filled road at a right angle, so that some bombs would be fairly certain to hit the target. In flying along the course of a road it often happened that all the bombs fell so far to the sides of the road that they were completely ineffective.

Bomber attacks at night bore the stamp of nuisance raids. In nearly all instances they were flown by individual planes and directed against targets located close to the front, such as billets, highway traffic, artillery positions, and other points occupied by troops. They were annoying to be sure, but seldom caused major damage. In order to make it difficult for the night bombers to locate their targets, especially on moonlit nights, the Germans smeared all whitewashed houses with mud and conducted all movements without lights. Installation of small Flak units at persistently attacked points caused heavy losses of Russian night planes, since their silhouettes in the moonlight made them easy targets. The attacks ceased immediately.

Night attacks directed against large German installations, such as railheads, airfields, factories, cities, etc., lasted all night. They were carried out by a chain of individual aircraft following each other at short intervals. Flights to and from the target followed different routes. Whenever the Russians aimed at destroying extensive installations, they repeated their bombing raids for several consecutive nights. In the summer of 1944, for example, most of the city of Tilsit in East Prussia was destroyed in a long series of Soviet night attacks.

On dark nights, Russian aircraft used parachute flares resembling Christmas trees to mark the direction of the bomb run and the target areas for the attacking bombers. Frequently, however, inhabited places thus bracketed by a rectangle of flares could be evacuated before the arrival of the first bombers. In other cases, the wind blew the rectangles so far away before the bombers appeared that all the bombs fell into open fields and caused no damage.

The Russians used German searchlights and other guide lights as navigation aids, and for that reason never attacked them. Russian night fighters seldom appeared and were of very minor importance.

A faulty sense of orientation in the air, or an inadequate knowledge of map reading, repeatedly caused Russian pilots accidentally to land behind the German lines. In January 1943, for instance, a Russian pilot accompanied by an engineer was supposed to repair another plane which had made a forced landing. Despite perfect visibility, the repair-shop plane strayed across the clearly recognizable river which formed the front and landed in German territory. Much more distressing to the Russians was the accidental landing in the German Buczacz bridgehead of a liaison plane with courier mail and an *Armeeintendant* (administrative official of an army) aboard. That incident took place shortly before the Russian summer offensive east of Lwow in 1944. The pilot was flying from Kolomea to Tarnopol, but landed at the halfway mark because the similarity of terrain and the brown uniforms of a Hungarian unit, which he mistook for Russians, had confused him.

The Russians repeatedly used captured German planes, with the German insignia retained, for purposes of reconnaissance and for occasional special missions.

PART FIVE
PARTISAN WARFARE

Chapter 16

Partisan Combat Methods

Generally speaking, Russian partisan groups on the Eastern Front were formed early in 1942. At first they were mainly isolated bands of little strength, frequently dropped from aircraft, operating in rear areas well behind the German front. During the summer of 1942, however, these bands were gradually combined into more closely knit groups, put under a unified command, and continuously reinforced. Accordingly, their operations grew in scope and impact.

Partisan group activities seldom covered areas near the front except when extensive, pathless forests favored their approach. In general, the partisan groups would maneuver in the rear areas of the German armies, in woods and swamps next to highways and railroads. They avoided open territory and regions occupied by German troops, but kept the latter under surveillance.

From the outset the German troops had difficulty defending themselves against this type of warfare. Its effectiveness had been underestimated. Apart from the fact that, considering the vast areas, the German forces were not numerous enough to combat the steadily expanding partisan groups, the front-line troops, which had been trained for orthodox warfare, all lacked experience in antipartisan warfare.

During large-scale enemy break-throughs, or German withdrawals, strong partisan groups frequently managed to coordinate their operations with those of Soviet cavalry, ski units, infiltrated infantry, or paratroops. Substantial German forces (usually several infantry and panzer divisions) had to be mustered in order to combat the joint enemy efforts. Prior to large-scale Russian offensives, strong bands would often migrate to the areas that the Red Army soon hoped to take. Such movements, therefore, gave some indication of Russian intentions. Prior to the beginning of the large-scale Red Army offensive in East Galicia in July 1944, for example, numerous bands worked their way into the Carpathian Mountains southwest of Lwow, which were among the objectives of the Soviet operations.

On the other hand, during each Russian withdrawal, as well as subsequent to battles of encirclement, innumerable soldiers cut off from their own forces, and sometimes entire combat units, made their way to the partisans and fought with them. In such instances, too, partisan activities developed into a serious threat.

During the winter, strong bands, well organized from a military standpoint and commanded by specially trained leaders, developed intense activity in the extensive woodlands of the Eastern Front.

The bands were generally organized into groupments of from 3,000 to 5,000 men each. As long as the front remained static, these groupments would remain in a fixed location; they were quartered in winterproofed camps, excellently constructed and heavily guarded. Smaller groups, varying greatly in strength, comprised at least 100 men. Attached to each groupment was a number of these smaller partisan groups. They branched out through the entire rear area and frequently were only in loose liaison with the groupment. They constantly changed their position and therefore were difficult to locate in the vast rear area, which was only sparsely occupied by German troops. They had contact men in all the larger villages of importance to them. Dispersed and cut-off Russian units gave them even tactical striking power. In 1941, for instance, in the area of Army Group Center, 10,000 men under General Kulik operated very skillfully and could not be cornered. Another example was the remnants of the 32d Kazak Division, whose destruction required the commitment of German front-line troops on 6 and 7 August 1941. In 1944, activities of partisans, reinforced by infiltrated troops, had reached such proportions west of the extremely swampy Narva River that the left wing of the northern front (III SS Panzer Corps) had to be pulled back in order to form a shorter and more easily guarded line.

Every camp of the larger partisan groups was secured on all sides—in some sections to a depth of several hundred yards—by thick underbrush, brier obstacles, or abatis and wire entanglements. All roads leading to the camp were blocked or camouflaged, or detours were built which led in another direction. Traffic to the camp was conducted on paths known only to the initiated. Sometimes these paths were protected by bodies of water, with crossings built 8 to 12 inches below water level, or by large stretches of swamp which could be crossed only on swamp skis. All movements of strangers were carefully controlled by sentries stationed far from camp and disguised as peasants. Strangers were also kept under close surveillance by a network of spies active in all villages in the vicinity.

The camps were well supplied with weapons, ammunition, explosives, and rations. Only very reliable partisans were put in charge of these supplies.

The camps procured their food supplies by forced requisitions in nearby villages. Villages refusing food contributions were ruthlessly put to the torch by the partisans; the men were dragged into the woods, and the women and children dispersed. Supplies were also received by aircraft, which dropped the rations in the immediate vicinity of the camp when prearranged light or fire signals were displayed. Looting vehicles during partisan raids likewise provided ammunition and small arms for the bands.

Excellent camouflage prevented any aerial observation of the camps. The shelters were allowed to be heated only at night, so that no smoke would disclose the existence of the camp during the day. The partisans succeeded in maintaining the secrecy of the camps for a long time by having small bands appear in remote villages and by disseminating false rumors concerning partisan movements. The mere suspicion of betrayal was sufficient cause for execution of the suspect. The same fate threatened the family of the condemned. These measures explain why all partisans operations were kept secret. Whoever joined a partisan group, voluntarily or involuntarily, could leave it only at the risk of his life.

The partisans also had signal communications at their disposal. The larger partisan units received their directives by short-wave radio, so that they had up-to-date information about current military developments in their respective sectors. Air couriers were also used. There was a carefully camouflaged landing place for liaison airplanes in the immediate vicinity of almost every major camp.

Practically without exception partisan operations were carried out at night. Daytime raids seldom took place, and then only in areas in which no German troops were stationed for miles around. Raids of that type were usually confined to individual motor vehicles.

A major partisan operation, with the demolition of a railroad bridge as its objective, would proceed as follows: A long column of women and children would move along the right of way in the direction of the bridge. Presuming them to be refugees, the German sentry would take no action. When the head of the column had reached the bridge, heavy surprise fire was directed against the bridgehead from the end of the column. Machine guns, set up on the roadbed in the direction of the bridge, pinned down the German guards. Under this fire cover, and by utilizing women and children in violation of international law, the partisans succeeded in installing prepared demolition charges and in destroying the bridge.

Partisan operations generally included mining main highways, demolition of railroad tracks, mining railroad beds and arming the mines with push and pull igniters, surprise fire attacks on trains, looting derailed railroad cars, raids on trucks and convoys, and burning ration, ammunition, and fuel depots. Less frequent were raids on command posts of higher German headquarters.

The partisans followed the practice of avoiding open combat as much as possible. This practice was indeed the guiding rule upon which their method of warfare was based. Unusual developments at the front would immediately result in extremely lively partisan activity, essentially aimed at the disruption and destruction of railroad lines. During a major German attack, for instance, the main line of a railroad that had to handle the supplies for three German armies was blasted at two thousand points in a single night and so effectively disrupted that all traffic was stalled for several days. Such large-scale operations, carried out by small partisan teams and numerous individuals, at times seriously hampered the supply of the German troops.

Chapter 17

Defense Against Partisan Activity

The German forces in Russia took both passive and active defense measures to protect their rear areas against the surge of partisan activities.

I. Passive Antipartisan Measures

Each army group created a special staff whose duty it was to collect all information concerning the appearance and movement of partisans by close contact with the military authorities in the rear areas and with Russian community heads, as well as by a network of agents in areas threatened by partisans. All information thus gathered would immediately be passed on to the German military authorities concerned.

Small headquarters were combined to protect them more effectively against partisan raids.

Local defense units were drawn from among the Russian population in threatened areas. Often Russian civilians urgently requested this measure because they suffered from confiscation of cattle, forceful removal of men, etc., by the partisans.

All traffic was halted on especially endangered roads at nightfall; such roads were used in daytime only at certain hours and in convoys escorted by armed guards.

Railroads, bridges, and trains were protected. Outguards within sight or earshot of each other were posted along railroad lines in threatened areas. The outguards were quartered in blockhouses protected by wire entanglements and abatis, behind which lay also the entrenchments for defense. Wherever the railroad line led through wooded terrain, all trees within 50 yards of either side of the right-of-way were felled to provide a better field of vision.

Reinforced outguards equipped with infantry heavy weapons protected all bridges. Another precaution, however, had to be taken in addition to furnishing local protection for bridges and adjacent bridgeheads, particularly whenever larger bridges were to be safeguarded. Strong guard detachments had to be posted at a great enough distance to permit them to spot approaching partisan bands, and to allow time for an orderly preparation of countermeasures. Precautions of that nature had not been taken prior to the previously mentioned partisan operation against the railroad bridge.

All trains going through danger zones had two sand-filled gondola cars coupled in front to protect their locomotives from mines. Each train was escorted by a guard detachment of about forty men.

Furlough trains were guarded by the soldier-passengers themselves. For that reason, all men going on furlough had to carry their small arms. Night traffic on particularly imperiled railroad lines was discontinued from time to time. During the day, the trains on these lines sometimes traveled within sight of each other. This procedure, however, was possible only because Russian air activity in the rear area was very limited.

II. Active Antipartisan Measures

Units such as security divisions and forces particularly organized for that purpose were normally assigned the mission of fighting the partisans. The great depth of area required a substantial number of such units, and since they were not available in desired numbers, security divisions frequently had to be assigned zones that they were hardly able to control.

In the forest terrain of Baranovichi and Minsk, for instance, the German 707th Security Division had to guard an area of 40,000 square miles (larger than all of Austria). Its duties usually consisted of protection of important points in seriously threatened wooded areas; surveillance and protection of zones and villages through which led military supply routes, and which were constantly imperiled by partisan bands; reconnaissance of partisan camps and roads leading to them; daily dispatch of as many combat patrols as possible into partisan territory to prevent the partisans from uniting into groups and establishing permanent bases; and operations against detected partisan camps.

Whenever the Germans planned a major operation against a detected partisan camp, the project had to be kept a strict secret from the troops. Experience had taught that if such plans were revealed, even larger partisan groups immediately dissolved, only to assemble again at a different location. It happened repeatedly that in carefully prepared operations partisan camps, which shortly before had been fully occupied, were found deserted. The troops, therefore, could only be informed of the actual plans after they had reached the outer line of encirclement.

The assembly of the attacking troops had to take place at least 1 day's march away from the partisan camps. The advance toward the outer line of encirclement had to be so timed that all troops could reach it simultaneously and occupy it immediately. As far as possible, the outer line of encirclement was anchored on natural obstacles that were easy to block and to keep under surveillance (for example, rivers).

The troops were deployed in the outer line of encirclement in such a manner that they formed a continuous line of sentries, with each soldier within calling distance—and at night, within sight—of the next. Behind this line of sentries, pursuit detachments were kept ready for immediate employment against partisan bands which might break out. As soon as the encirclement had been completed, leaflets were dropped over all inhabited places within the ring, ordering all inhabitants to evacuate at once and to asesmble at a designated point.

The contraction of the ring of encirclement proceeded during daytime only, in phases covering not more than 2 to 3 miles per day, and the territory was carefully combed. Individual sectors had to be occupied by at least 2 hours before twilight, so that the troops could establish themselves and become acquainted with the terrain ahead while it was still light.

Sectors easily distinguishable in the woods (glades, paths, railroad lines) were designated as the new line of encirclement. Close contact between individuals had to be maintained. Nighttime security at the sector boundaries was of particular importance. The procedure of detailing forces for guarding unit boundaries, as well as the command over those forces, had to be clearly regulated. The further contraction of the ring up to the final encirclement of the camp followed the same pattern as described above.

As soon as the encirclement had started, the surrounded area was kept under constant aerial observation. By dropping messages, the planes immediately notified troops and officers in command of any observed breakout attempts. Since breakouts were to be expected mainly at night, sufficient security detachments had to be posted in front of the sentry line. With the contraction of the ring of encirclement a proportionate number of reserves could be withdrawn, and their follow-up had to be properly regulated. If the partisans still remained in their camp by the time the troops had reached the final line of encirclement, a heavy air attack would usually enable the troops to score a quick success.

Experience had taught the Germans that this type of antipartisan warfare, though requiring large numbers of troops and much time, promised the greatest success. No other methods proved themselves in wooded terrain, since breakouts at night could hardly be prevented. Rigid discipline was a prerequisite for the success of such an operation. The designated objective for the day could not be changed during the operation, and the slightest independent changes on the part of the troops would disrupt the line of encirclement and make the breakout of partisans possible.

Winter proved to be the most favorable season for antipartisan operations because all movements could be more readily observed in snow-covered terrain. In summer the dense foliage of the woods made such operations very difficult. As far as possible they were to be carried out during bright nights, best of all during a full moon. Liberal armament with machine pistols proved advantageous. Mortars were found to have more of a demoralizing than actual effect, since their shells burst in the trees. Artillery could hardly be used during advances in woods. As a rule it could be put into action only during the battle for the fortified camp itself. Depending upon the terrain, the Germans found it advisable to have individual guns follow directly behind the leading elements. The employment of tanks, where the terrain was suitable, produced excellent results. In such operations the troops had to have an adequate supply of signal pistols and cartridges. In the case of swampland, the troops were to be equipped with swamp skis.

The wooded terrain, which afforded poor visibility, and deceptions at night often caused shooting frays that started a panic among German units or resulted in their firing on their own troops. The Germans therefore found it advisable to prohibit the firing of all infantry light weapons except during partisan attacks. Special regulations for opening fire also were required when the final ring of encirclement had been closed and the troops were facing each other at a short distance.

Chapter 18

Non-Russian Partisans

In addition to the Russian partisan groups, there also existed in the East strong Ukrainian and Polish groups, as well as a few weak Czech and Jewish groups. The latter two were of no great importance. Some of the bands were for, and others against, Russia. They fought each other cruelly and ruthlessly to the point of annihilation. In 1944, for instance, at the Polish-Ukrainian linguistic frontier, Polish bands raided Ukrainian villages, and Ukrainian bands raided Polish villages, burned them, and massacring the entire populations including women and children. There were insufficient German troops to occupy the entire territory densely enough to prevent such raids. Emergency detachments usually came too late.

Behind the German front, severe fighting, even to the extent of employing heavy weapons, frequently broke out between the partisan groups of different camps. Such disturbances at times caused a local paralysis of Soviet partisan activity.

The very active bands of the Ukrainian Nationalist Movement (UPA) formed the strongest partisan group in the East except for the Russian Communist bands, which they fought bitterly. The UPA repeatedly offered its cooperation against the Soviet partisan bands to the German Army and asked for a German general to act as organizer and tactical leader, but the High Command of the German Army, very much to its disadvantage, continually refused these requests. Only tacit, local agreements were therefore made between the UPA and German military authorities. They proved advantageous to both parties. Sometimes the UPA would participate in fighting the Soviet bands even without any previous agreement (as in 1944 north of Lwow and at Stanislav). The UPA forces were deployed in groups of several thousands each in the rear of the Russian, as well as the German, armies. Although they fought the German political organizations and their police forces, they never fought the German troops, and they seriously harassed the Soviet Army. Not only did they severely impede its supply, they also attacked Russian headquarters and rendered them powerless by encirclement (for example, in 1944 at a Russian corps headquarters in Korosten, which had to be liberated by motorized troops). Furthermore, they incited revolts of the Ukrainian population in the rear area (in Kiev, for example), the quelling of which required the withdrawal of several Russian divisions from the front for an extended period of time.

Although seriously disturbing German supply operations, and thereby also the conduct of battle, even the strongest partisan movement, that of the Soviets, had little influence on the over-all operations in the East. This is illustrated in the report on the protection of supply lines near Kirovograd against partisan raids (1942–43).

The Kirovograd region was administered by German civilian authorities, with the aid of municipal and regional commissioners who had their seats in the larger towns. The former town and community boundaries were retained as much as possible. The security of the region was the responsibility of the regional military government officer (*Feldkommandant*) and the local military government officers (*Ortskommandant*) under him. It included mainly the security of railroads, the two important supply routes (express highways IV and IVc), and the Bug bridges at Pervomaisk and Voznesensk. Two local defense (*Landesschuetzen*) battalions were available for these tasks.

Except for occasional minor sabotage acts against railroads, telephone lines, etc., by the native population, the area administered by German Regional Military Government Office (*Feldkommandantur*) 509 remained absolutely undisturbed throughout 1942, and substantial progress was made in the work on express highways IV and IVc. Not until March of 1943 did a sizable partisan group invade this area.

Of the railroad lines in this region, the one with the branch line to Dnepropetrovsk and Krivoi Rog, via Aleksandriya–Pyatikhatka, was the most vital from the military as well as the war-economy standpoint. The important junction and marshalling center of Znemenka was the point most vulnerable to sabotage operations. Had that place been destroyed, the entire supply of the front, and shipments of badly needed war materials destined for Germany, would have become seriously endangered for a rather long time. The terrain was level and flat, with little vegetation. It was partially steppeland. Except for a fairly large, dense patch of forest between Znemenka and Tsibulnos, only smaller strips of woodland limited the visibility along the railroad line. Features extremely favorable for sabotage operations were the extensive fields of sunflowers, growing taller than a man, and to a lesser degree, the grain fields.

Raids on the railroad line were always carried out in insidiously cunning ways by bands which found refuge mainly in the woods north of Znemenka and which were willingly aided by elements among the population of the neighboring villages. Organized into small combat patrols (*Troikas*), and carefully protecting their lines of withdrawal, they blew up railroad tracks, damaged control towers and signal installations, and terrorized the native railroad personnel. The *Troikas* disappeared upon the completion of their work of destruction, as a rule without having suffered any losses. Their activities were directed from a higher level, and they were regularly supplied with weapons, communications equipment, explosives, maps, medical supplies, etc., which sometimes were dropped by aircraft. The explosives consisted of demolition charges, mines, artillery shells, or improvised infernal machines.

The weapons of the *Troikas* were generally machine pistols, daggers, and rifles. The members wore civilian clothes, but some were dressed in uniforms of the German Army, Luftwaffe, or political organizations.

In March 1943, a partisan group, at first numbering from 200 to 300 men, with a few women as medical and signal personnel, invaded the region of Feldkommandantur 509 at Chigirin (35 miles west of Kremenchug). It was very ably led and highly mobile in its conduct of battle. The group used sleighs in the winter and requisitioned horses during the muddy period. For 3 weeks they roamed through this region; raided agricultural centers, sawmills, and other plants; killed the managers of the agricultural centers as well as the German harvest control officers; liberated prisoners of war; and disrupted rail traffic at Sosnovka. The partisans sought shelter in villages adjacent to woodlands. Horses and vehicles were sheltered in steep ravines to project them against artillery fire. The expertly arranged security and reconnaissance measures precluded the possibility of surprise raids by German troops. Orders and instructions were received by short-wave equipment which was kept in a suitcase. Because of their well-functioning intelligence machinery, searches of woodlands and villages for the smaller sabotage teams of partisans were almost never successful. The Ukrainian community authorities' lists of inhabitants were seldom in order, making identification of the inhabitants of a village very difficult. It was, therefore, found practicable to chalk the names of the residents of each house on the front door.

In combatting the partisans, railroad stations, control towers, wooded zones, and bridges were focal points of security. Shifting German reserves by motor vehicles was planned for a quick reinforcement of the defenses. All places that had to be defended were converted into strong points. Arrangements had to be made in buildings to permit firing from various floors, and machine gun emplacements with flanking and alternate firing positions had to be installed. Plank walls had to be erected for protection against fragmentation, and windows had to be provided with gun shields and embrasures. Fire steps, ammunition lockers, communication corridors, and observation posts had to be established. Telephone lines were laid so as to be protected as much as possible against gun fire, and supplemented by optic and acoustic alarm signals. In the vicinity of each German strong point, the terrain was cleared of obstructing vegetation for a distance of about 1,000 yards in each direction along the railroad line; this cleared area extended some 300 to 400 yards from the tracks.

The goal of all these precautions and preparations was to assure the smooth functioning of rail traffic. In addition to military security, the railroad lines were frequently inspected by employees walking along the tracks, and patrolled by handcars, or locomotives pushing freight cars ahead of them. The speedy formation of "alert" units for fire fighting, and the procedure of transporting them by motor or rail had to be planned and practiced in drills. The use of armored trains, or the addition to trains of cars carrying troops, proved to be effective methods of antipartisan defense.

The Germans soon realized that their military forces were not sufficient for the manifold tasks of railroad security. To ease their burden, local inhabitants were assigned to the troops as so-called voice-alarm sentries. Upon spotting partisans, these sentries were to shout the information to their neighboring guard posts, stationed at a distance of five hundred yards, but were not to fight. For that reason the native sentries were unarmed. Their reliability varied, since they feared the cruel reprisals of the partisans against themselves and their families. Organized indigenous forces, some of which were employed for railroad defense, likewise functioned only when they were heavily intermixed with German units. The German regional defense units were not properly armed. Their equipment, too, was completely inadequate. For reconnaissance, the *Fieseler-Storch* (liaison aircraft) proved equal to all demands. The evacuation of villages directly adjacent to railroad lines, and the establishment of so-called death zones, within which any civilian would be shot on sight, were found to be appropriate measures.

PART SIX

CONCLUSIONS

The success of any armed force flows from a sum total of measures not only in the military field, but also in politics, national indoctrination, economics, and other spheres, which in their entirety decide the outcome of a war. They all must therefore be analyzed and weighed against each other if one would find the ultimate causes of victory or defeat. Only the product of intensive studies would form a positive basis for the evaluation of an armed force in a past war, and thereby offer clues as to its potential in a possible future conflict.

In the preceding pages, only the military measures—in themselves important—could be taken under scrutiny. But this study, presenting German experiences as well as German views, throws light on many significant aspects that permit readily applicable conclusions to be drawn.

Not the success alone, but also the circumstances under which it was achieved form the standards by which an armed force is to be evaluated. The Red Army was successful in the last war, a good argument for the proposition that in a future war, fought under equal or similar circumstances, it would again emerge victorious. The high command was good and in its hands the troops, purely as a human mass, were a useful instrument. The prime motive force behind both was communism; the final goal, World Revolution.

Manpower and matériel were abundant, and may be presumed to be abundant in the future. The Soviets gained a wealth of war experience on the basis of which to train their leaders and masses in up-to-date methods. They also have the means to maintain quantity and quality of their equipment at a high level. Accordingly, one might think the final conclusion from this study and from the description of climatic conditions in the U.S.S.R. has to be, "Hands off Russia; the Russian combination of mass and space cannot be overpowered!"

That, however, is not the case. It should not be forgotten that much that might be said about Russia and the Russian had to be condensed here as well as in other studies. The terrain difficulties and the characteristics of the Russian soldier had to be described one after the other in such a way that both could easily be overestimated.

Despite Russia and the Russian, despite cold and mud, despite inadequate equipment and a virtually ridiculous numerical inferiority, the German soldier actually had a victory over the Soviets within his grasp.

The prerequisite for a successful war against the U.S.S.R. is a systematic preparation for the undertaking. One cannot provoke such a conflict and expect to carry it through in a spirit of adventure. The equipment of the soldiers and the total amount of matériel must meet the requirements of Russian terrain at all seasons. This is a question of industrial potential which, by applying the experiences gathered in the last war, is not difficult to solve.

Training, tactical as well as physical, must comply with the abovementioned peculiarities of combat against Russia and on Russian soil. Training aimed at imparting toughness, independence, and willingness to assume responsibility, and the molding of self-reliant individual fighters, as well as leaders who are willing to take chances, are the most essential points in this respect. Strict discipline is an additional fundamental condition for every one who fights in Russia. Even the best athletic background is insufficient to meet the severity of this test. And, last but not least, the soldier must have an inner conviction—the indomitable will to prevent the Russian Moloch from devouring the world.

☆ U.S. GPO 415-394/1983

Map I
REFERENCE MAP
RUSSIAN COMBAT METHODS
SCALE
100
100
200
MILES
SWEDEN
FINLAND
GULF OF BOTHNIA
WHITE SEA
ZASHEYEK
KANDALAKSHA
Lake Ladoga
GULF OF FINLAND
LENINGRAD
POSYOLOK TAYTSY
KINGISEPP
KRASNOGVARDEYSK
PORECHYE
BALTIC SEA
ESTONIA
LATVIA
LITHUANIA
RASEINIAI
EAST PRUSSIA
U S S R
Volga R
MOSCOW
VYAZMA
KOLOMNA
Oka R
KASHIRA
RYAZAN
TULA
MIKHAYLOV
STALINOGORSK
MINSK
MOGILEV
Dnepr R
SLONIM
SLUTSK
BOBRUYSK
ROGACHEV
DYATKOVO
BRYANSK
KARACHEV
POCHEP
Vistula
WARSAW
PRUZANA
KOBRYN
BREST LITOVSK
POLAND
KOWEL
RECHITSA
GOMEL
PRIPYAT
MARSHES
YELETS
Desna R
BYELGOROD
BOGODUKHOV
KHARKOV
Don
River
Volga River
Dnepr
KALACH
STALINGRAD
Donets R
POTEMKINSKAYA
ZIMOVNIKI
SEA OF AZOV
CZECHOSLOVAKIA
RUMANIA

www.ingramcontent.com/pod-product-compliance
Ingram Content Group UK Ltd.
Pitfield, Milton Keynes, MK11 3LW, UK
UKHW020610050125
453105UK00013B/76